KT-503-761

HOW TO LIVE
FOREVER

AND 34 OTHER
REALLY INTERESTING
USES OF
SCIENCE

ALOK JHA

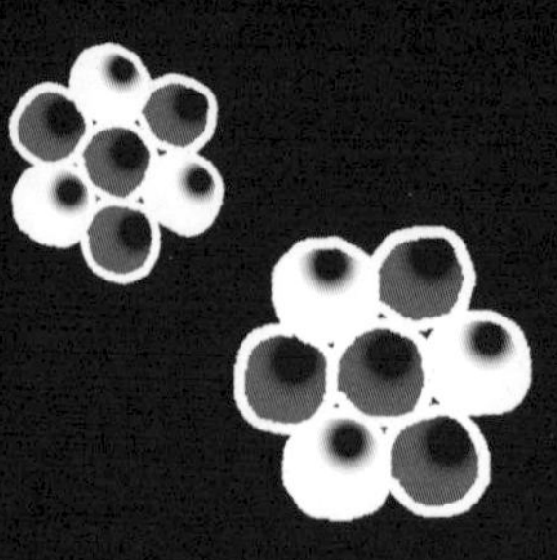

Contents

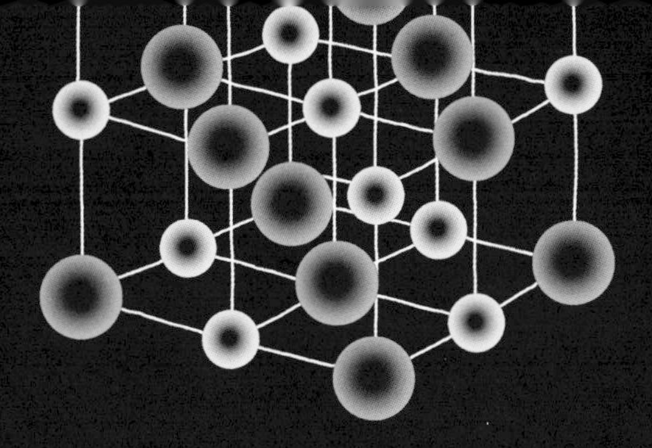

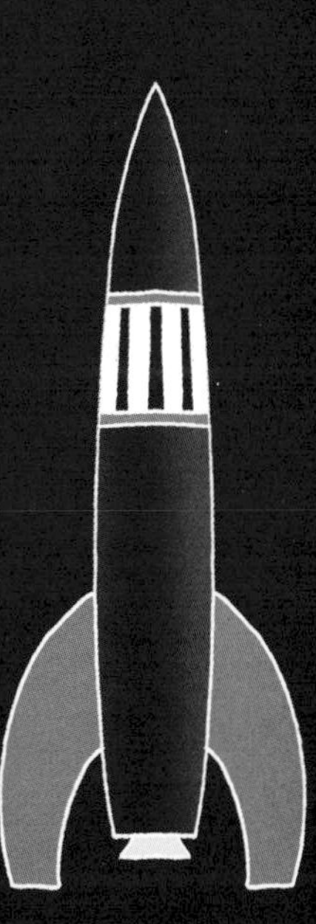

Introduction

How do you work out the pressure of a gas on the side of a container? First, imagine a single molecule of gas flying around a container and calculate how hard it hits the surfaces. Scale that model up to work out the combined force per unit area, or pressure, of a system of trillions and trillions of molecules. It's so simple 15-year-olds could do it. Don't believe me? Well, I was one of them.

On a winter morning 18 years ago, I finally understood what science was. Until then, it had been a black box of complex laws and descriptions of how things worked. I was convinced the laws of nature were dug out of the ground like mud-encrusted fossils. (I never worked out how. Did they go on digs like archaeologists? If so, where?) And they had all been found anyway. There was no more work to do, just a list of things to learn so that I could pass some exams.

It wasn't that I had no interest in the world around me. I read books about how stars were formed, made newspapers burn with lenses, collected insects and tried mixing household chemicals, like any child might. But in my mind, none of that curiosity related to what was going on at school. Every physics, biology and chemistry lesson was simply a case of watching another equation, law or definition drop out of a black box.

Salvation for me came from my high school physics teacher. He would plough through the minutiae of the syllabus, but he also taught us what science actually is. Often, he would just set a problem – work out how hot the surface of the Sun is, given the temperature of Earth's atmosphere, say – and then walk out of the classroom, asking one of us to find him when someone had worked it out. He indulged our curiosity. A classmate was convinced, for example, that crushing a Polo mint gave out a faint blue light, so our teacher gathered us in the photography club's darkroom to see if it was true. We didn't prove it that day but it stoked our curiosity for strange questions, which he would patiently answer.

In those two years, we were asked to do what Robert Boyle, Joseph Gay-Lussac and Jacques Charles did from first principles nearly 200 years earlier – work out how pressure, temperature and volume were related. We even had a go at working out how Albert Einstein came up with the idea of the photoelectric effect, which marked the birth of quantum theory. And the thing was, we could do it. These giants of science, immortalized in the names given to physical constants and the laws of nature, had spent decades arriving at their answers. But we could follow their logic. Even better, working out those equations or the answers to apparent conundrums gave a sense of ownership. The black box was beginning to crack open and the decisive move had been made by my own imagination.

Science is all about solving problems and, in many cases, those problems revolve around framing the right question to ask. Perhaps you want to know why nuclear bombs are so powerful or what happened in the second after the Big Bang. Maybe you've spent a night or two looking up at the heavens and wondering if there is life out there, or just been curious about how electricity gets into the plug sockets at home. Or you want to know whether or not to worry about the greenhouse effect or how the rich diversity of life has evolved on Earth over its 4.5 billion year history. Maybe you want to know how we found out that Earth is 4.5 billion years old?

These questions (and so many more) have all been answered by one of the most remarkable, creative and collective efforts in the history of humanity. Science has given us the tools to unlock some of the most profound mysteries of the Universe.

I was enraptured by science at age 15 because I remember thinking that there had to be more to the numbers and rules. But plenty of others probably leave behind what they think is a boring and difficult subject at school, forever more carrying an impression that it's reserved for the nerdy. For anyone who still harbours such thoughts, I hope that you will find in this book a glimpse into what science really is: the remarkable story of the human imagination.

1 How to clone a sheep

- It starts with a cell
- The difficult biology of cloning
- After Dolly
- Clones to the rescue

In 1997, a very ordinary black-faced mountain sheep gave birth to a very extraordinary lamb. Improbably for a lamb, her arrival generated tens of thousands of words in newspapers, grabbed the gaze of scientists from around the world and induced a bout of soul-searching that still continues today. The birth of Dolly the sheep kick-started a new era of science and, along with it, a whole new set of moral questions for society to grapple with. The reason? Dolly had been cloned from another adult sheep.

It starts with a cell

On paper, cloning is simple. Take a cell from the individual you want to clone and extract the DNA. Put that material into an unfertilized egg, which has had its own DNA removed. Trick this composite egg into dividing, usually with a jolt of electricity, and let it grow in the lab for a few hours or days. If that works, transfer the dividing embryo into a surrogate womb and keep your fingers crossed that the embryo becomes a baby.

In real life the process of 'somatic cell nuclear transfer' (SCNT), as the standard cloning technique is known, is a lot harder than it sounds. The scientists at the Roslin Institute in Edinburgh, who created Dolly the sheep, and colleagues in molecular biology laboratories around the world, had been working for decades to understand the SCNT technique before their 1997 success. They carefully watched everything from frogs to mice, prodding and teasing apart embryo cells at various stages to work out how a single cell manages to grow into a complete organism.

'We now report the birth of live lambs from three new cell populations established from adult mammary gland, fetus and embryo.'

IAN WILMUT

Molecular biology has its humble origins in the gardens of an Austrian monk called Gregor Mendel. In the 1800s, he became the first person to study heredity in a systematic way by tracking how the various visible traits of peas, such as the colour of the flowers or whether the peas were smooth or wrinkled, changed over successive generations of plants. He postulated that visible traits seemed to have 'factors' of transmission associated with them, which parent plants passed on to their offspring in seeds. Mendel did not know what these factors were but, by the start of the 20th century, scientists began to gather evidence that the transmission had something to do with the DNA at the heart of cells.

How this molecule passed on the information was revealed when, in 1952, Jim Watson and Francis Crick, working in the UK, proposed a structure for DNA: a double helix molecule with a sequence of nucleotide bases,

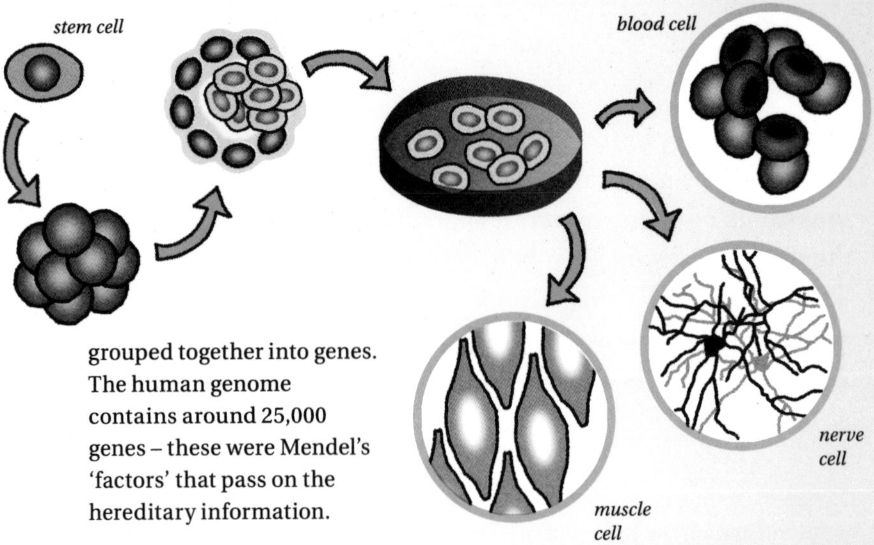

stem cell

blood cell

nerve cell

muscle cell

Stem cells, especially those at the centre of a growing embryo, can turn into any cell in the body, such as blood, muscle or nerve cells.

grouped together into genes. The human genome contains around 25,000 genes – these were Mendel's 'factors' that pass on the hereditary information.

Every physical (and, in animals, some non-physical) characteristic is passed from parent to child in the information encoded in DNA. Undoubtedly the most famous molecule in science, the long strands of the double helix are housed in the nucleus of living cells. They contain a precise sequence of four nucleotide bases – C, G, A and T – which give the machinery inside cells the instructions they need to create every protein the organism will need. Everything from the chemical signals that regulate the everyday functions of life to the physical materials that make up muscles, leaves and bones comes from the instructions contained in an organism's DNA. In an animal, every cell (apart from the egg and sperm) contains a full complement of DNA, although not all parts of the full genome are active in all cells.

As an organism grows, different parts of the genome become activated in different parts of the body at different times. As a boy undergoes puberty, for example, instructions from the brain (in the form of hormones) go out around the body to make the voice box lengthen, muscle mass increase and for hair to grow in new places.

The difficult biology of cloning

Normal sexual reproduction involves creating offspring by mixing half of one organism's DNA with half from another. In everyday terms, that means a child is produced from combining half of the mother's DNA and half of the father's. Cloning bypasses this mixing and takes the child's DNA from a single parent. The first animal copy made in this way was created by the godfather of cloning, German embryologist and Nobel prize winner Hans Spemann, in 1938. He cloned salamanders using nuclear manipulation techniques, and also came up with the idea for SCNT, although he did not have the technical capability to carry it out at the time.

SCNT was eventually carried out successfully in 1952 on American leopard frogs. Mice and rabbits were followed by the first cloned sheep in 1986, made from early-stage embryos by Danish scientist Steen Willadsen, who would later work with Ian Wilmut, a British embryologist nicknamed the Father of Dolly, to create the headlines of 1997. To make Dolly, the Roslin Institute scientists, through a large amount of trial and error, worked out that the donor DNA for a cloning procedure had to come from a cell that was at the very start of its cell-division cycle. This DNA was inserted into the empty egg using a thin glass needle, and an electrical shock jogged the reconstructed egg into a stage where it could start dividing.

The process is notoriously inefficient. During early experiments on sheep, scientists wanted to test whether the reconstructed eggs would work at all, so they took the DNA out of 244 naturally fertilized embryos and implanted each set into a new hollowed-out egg. Only 34 of these eggs developed to a point where they could be implanted into a womb and, from those, only five animals were born in June 1995. Three died soon after birth but two survived, and were named Megan and Morag.

Having proved that the technology could work, the Roslin scientists attempted full-scale cloning using donor DNA from adult, rather than embryo, cells. Early-stage embryos contain stem cells that are not yet specialized. In other words they have not yet received the instructions to turn into, for example, heart, liver, bone or muscle cells. Growing all the cells required to build an entire organism from these embryonic stem cells, therefore, does not seem to be too much of a stretch.

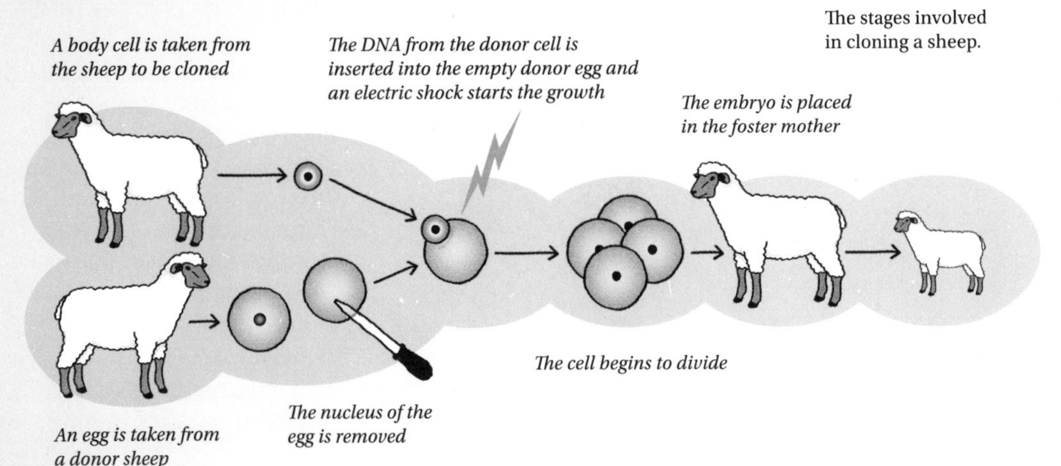

The stages involved in cloning a sheep.

But what about using mature, specialized cells from an adult animal? In a skin, bone or heart cell, for example, a lot of the DNA lies dormant because the instructions they encode are not needed for that cell's daily functions. Could the DNA inside such mature cells be regressed somehow so that all of this dormant genetic material is reactivated? The Roslin Institute's work culminated, after 277 attempts, in the creation of Dolly, a lamb grown from the DNA of an adult breast cell. (In a demonstration that scientists at the very edge of knowledge still know how to raise an eyebrow in humour, they named the sheep after US country-and-western singer Dolly Parton.)

After Dolly

Once Dolly showed that mammals could be cloned from adult cells, the race was on to reproduce the SCNT technique in other animals. So far, sheep, goats, dogs, horses, cows, mice, pigs, cats, rabbits and a gaur (a type of wild cattle) have all been cloned. In 2007, scientists created the first cloned primate embryo when they copied a macaque monkey.

Humans have proved somewhat more difficult to clone. It took almost another decade after Dolly before human embryos were successfully cloned, but none have yet survived for very long. In any case, the ethical issues around human cloning have led many countries to ban the technology – the United Nations adopted a non-binding resolution in 2005 calling for a worldwide ban on human cloning. Ask scientists, however, and they will tell you that creating new people using cloning has never been the point of research. So-called 'reproductive cloning' of humans will be technically difficult, if it not impossible – Ian Wilmut, writing in his book *After Dolly*, is unsure whether we will ever understand enough to flawlessly clone a person.

The real reason cloning interests researchers is its medical potential. No serious scientist is suggesting that we would need to produce live clones. Instead, after the SCNT procedure, the egg would be allowed to grow for a few days until it produced stem cells, after which the cloned embryo would be destroyed. This technique, technically identical to the first few days of reproductive cloning, is called 'therapeutic cloning'.

'...and these new clones are great for conquering green pastures.'

Clones to the rescue

Therapeutic cloning raises hopes for treatments that are well beyond anything possible today – generating custom-made tissues and organs for transplants, for example. Stem cell treatments might start by taking a sample of DNA from a patient and inserting it into a hollowed-out egg. When this reconstructed egg starts to divide, its embryonic stem cells would be harvested after a few days. These stem cells, genetically identical to the patient, could then be used to generate whatever organ or tissue is required, or to replace brain cells in intractable conditions such as Alzheimer's or Parkinson's. Such transplants would not be rejected by the patient and would drastically cut the number of donated organs needed in hospitals.

Scientists have had some stunning successes – renowned Anglo-Egyptian cardiac surgeon Sir Magdi Yacoub of Imperial College, London, has created parts of a human heart from stem cells. But there's a lot of research still to do before therapeutic cloning is anywhere near ready for use in the clinic. Extracting usable stem cells from human embryos is difficult and not without its objectors – many religious groups take issue with the idea that an embryo, a human life in their eyes, has to be destroyed in order to do the work. Stem cell research also contains a path yet to be trodden – knowledge of the complex chemical and environmental cues needed to turn stem cells into specific body cells is in its infancy.

There are few bounds to possibility if cloning research can be done properly and openly. One idea on the edges of possibility (ethically and technically) is the genetic modification of animals to produce 'humanized' versions of organs. This idea, called xenotransplantation, is an active area of research. Of all the animals that have been cloned successfully, the closest tissue match to humans are pigs (primates would be a closer genetic match but they are much harder to clone and take longer to reproduce). The idea with pigs is to create clones that are genetically altered to knock out the genes that might cause the human immune system to reject a transplanted organ. British scientists have been making headway in this area, announcing in recent years that they have managed to damp down the action of two of the genes active in pig tissue that might cause a person to reject their organs. Some researchers are even optimistic of human trials within a decade.

But, like all of the possibilities in the field of cloning, each use will raise a new slew of ethical questions. Should a pig die so you can live? More importantly, would you want a pig's heart beating in your chest?

2 How to start a plague

- How epidemics start
- The germ theory of disease
- Different types of bugs
- The machinery of germs
- How the body responds
- Preparing for an attack
- The microbes fight back

In the film* Twelve Monkeys, *a lone terrorist wipes out most of the world's population with just a box of vials that he carries onto an aeroplane. Each vial contains pathogens that will spread quickly around the world as infected passengers make their way to their destinations. It's an alarming, but very credible, plot line. At a time when it is possible to travel anywhere in the world in less than a day, it is also possible to transmit unpleasant things that far and that quickly too. Where humans go, so do the bugs that can injure or kill us.

How epidemics start

Before aeroplanes shrank the world, a disease outbreak in one part of the world would have had natural barriers to its spread. Anyone infected would either get better or die without moving very far from the place where they picked up the germ. They would have come into contact with relatively few people and avoided passing on their bugs. After a while, if quarantine procedures were tight, the germs would have nowhere else to go and no-one else to infect, and would disappear. The outbreak would be contained.

Nowadays international plagues are a big risk. Someone infected with influenza could board a plane in Beijing before showing any symptoms, arrive in London 12 hours later, see family or friends and then, only two or three days later, realize he has a temperature. In the meantime, he has exposed hundreds of people on the aeroplane, staff and passengers at the airport and also his friends. Some of those people will become infected and carry the germ to other far-flung parts of the world. Hundreds of infections quickly becomes thousands, tens of thousands and then millions.

The germ theory of disease

The realization that germs (everything including bacteria, viruses and parasites) cause disease is probably biology's single most important contribution to public health. The word germ, which comes from the Latin for 'seed', was first used in relation to disease in 1546 by Italian physician Girolamo Fracastoro. The first inkling of what germs looked like (or what they were) had to wait until the development of the microscope in the early 17th century, when Dutchman Anton van Leeuwenhoek identified what he called 'animalacules' in water and on human teeth – tiny objects swimming around, unnoticed until then. As microscopes improved, so did imagery, and understanding, of

A typical bacterium is a fully functioning cell. DNA in the centre is surrounded by a cell wall and the bacterium could also have a tail (flagellum) to help it move.

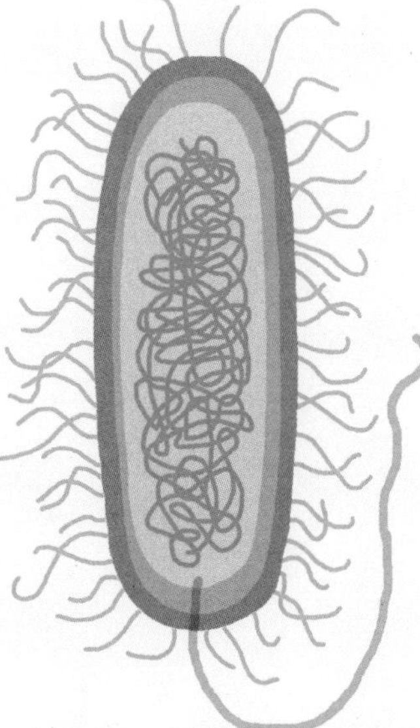

van Leeuwenhoek's animalacules. In 1835, Italian biologist Agostino Maria Bassi showed that a fungus caused muscardine in silkworms. In 1850, French physician Casimir Joseph Davaine found rod-shaped structures that he called 'bacteridia' in the blood of animals that had died of anthrax.

Though the evidence grew, germ theory was not easily accepted by doctors. If it seems obvious now that microbes cause disease, spare a thought for the Hungarian obstetrician Ignaz Semmelweis. Working in Vienna's Allgemeines Krankenhaus ('General Hospital') in 1847, he noticed that women who gave birth with the help of doctors or medical students were more than twice as likely to develop puerperal fever, also known as childbed fever, than women who were helped in labour by midwives. This disease was common in hospitals in the mid 19th century and could be fatal, with a mortality rate of up to 35 percent. Semmelweis noticed that the fever was especially common when the attending doctor had come directly from an autopsy. He put forward the idea that puerperal fever was an infectious disease and that it had something to do with the dead bodies in autopsies. As such, he encouraged doctors to wash their hands in chlorine solution after autopsies, a simple procedure that reduced the mortality from childbirth ten-fold at his hospital. The wider medical establishment of the time, however, was dubious about the germ idea. They rejected Semmelweis's thesis as he got ever more desperate to spread his message. He wrote open letters to prominent doctors, calling them murderers, and even his wife thought he was losing his mind at one point. Semmelweis was committed to an asylum in 1865, where he died shortly afterwards of septicaemia. It would be decades before his ideas were accepted.

An influenza virus, showing RNA material (at the centre) surrounded by a coat of proteins.

Despite the work of Semmelweis, hospitals remained largely unsanitary places until French chemist Louis Pasteur postulated three ways that germs might be destroyed – heat, chemicals or filtration. The British surgeon Joseph Lister experimented with antiseptic treatments for wounds using carbolic acid and also sprayed surgical instruments with the substance, attempting to prevent bacteria from getting into wounds in the first place. Both techniques reduced the instance of gangrene in Lister's patients and the techniques of antiseptic surgery became widely adopted by the close of the 19th century. Semmelweis was posthumously elevated to his rightful position as an early pioneer of antiseptic techniques, as germ theory took hold.

Different types of bugs

Even as germ theory started to become mainstream with the acceptance of the existence of bacteria, scientists began to wonder if bacteria told the whole story. In 1896, Dutch biologist Martinus Beijerinck proposed that water-soluble microbes, too small to be seen by microscopes, might be another way to transmit disease. These would pass through the filters that had been developed to stop bacteria. When scientists found that whatever caused tobacco mosaic disease, for example, could pass through filters, Beijerinck named these new particles 'filterable viruses'. He wondered if some sort of contagious living fluid was the cause. Filterable viruses were soon found to be at the cause of diseases such as influenza, rabies, vaccinia, foot and mouth, yellow fever and herpes simplex.

'Fortune favours the prepared mind.'

LOUIS PASTEUR

Today we know that there is no living fluid – we simply call these particles viruses – but when germ theory was being developed, the word *virus* (Latin for poison or morbid principle) was a generic term for anything that caused or transmitted disease. However, by the 1930s virus was a distinct term from bacteria. Imagery of viruses had to wait until the early 1930s and the development of the electron microscope, which revealed that viruses had regular structures and came in an array of shapes and sizes – from simple rods to intricate structures with tails. It also showed that they are not picky about what to infect – French-Canadian biologist Félix d'Herelle discovered bacteriophages, which are viruses that infect bacteria.

The machinery of germs

Unlike bacteria, which are complete cells that can exist by themselves as long as they have a source of nutrition, viruses parasitize their host and are unable to replicate by themselves. Usually composed of a protein coat encasing a strand of some genetic material, there is some debate about whether they can actually be classed as alive. When a virus infects a cell, it injects its DNA and the cell's machinery ends up doing what it always does when it comes across this sort of molecule – it replicates it. As the cell makes more copies of the virus, it is diverted from its real job of keeping its own body alive. Eventually, after the cell is filled up with copies of the virus, the copies burst out and go on to infect more cells. During this process, known as 'lysis', the body cell dies. As more and more cells die, the infected organism begins to suffer and the symptoms manifest themselves as disease. The symptoms of most virus infections are systemic – influenza, for example, produces runny noses, coughs and body aches. Bacterial infection symptoms are more localized – an infected cut, for example, will be painful around the site of the wound, while a bacterial throat infection usually sits on one side of the throat. However, some bacteria, such as *Clostridium botulinum*, secrete toxins that can cause damage around the body, such as muscle paralysis.

How the body responds

Given all their myriad tricks, why have all these minute particles not wiped us out completely? Fortunately for most multicellular organisms (including us), foreign invaders are repelled by the immune system. In humans, the immune system is composed of white blood cells that patrol the body in search of anything that looks foreign. Different types of white blood cell do different things – some of them engulf invading particles, while others produce antibodies that can kill germs. Once the immune system has killed off the infection from a particular germ, it will remember how to tackle the bug in future, as the relevant antibodies will continue to circulate in the bloodstream.

However, the immune system can be a double-edged sword. Often it is a powerful immune response to a germ that causes the symptoms of an ailment, such as high fever and inflammation, and sometimes these can be more devastating than any direct damage from the microbe. The victims of the deadliest flu pandemic in history, the 1918 Spanish flu, were killed when their bodies unleashed an uncontrolled immune reaction as a protective mechanism. Patients' lungs rapidly became inflamed and filled with blood and other fluids, which eventually drowned them. That strain of influenza ravaged populations around the world, killing an estimated 50 million people before it eventually died out.

Preparing for an attack

The immune system can be trained even before a person gets infected with a microbe. In 1800, English scientist Edward Jenner developed the first vaccination when he confirmed previous anecdotal evidence that prior infection with a benign disease of cows, called cowpox, protected humans from the more lethal smallpox. Louis Pasteur extended this technique to other diseases, both for bacteria and viruses. He weakened the fowl-cholera virus to reduce its virulence and also developed vaccines for anthrax and rabies. Weakened pathogens are now regularly used to induce a mild illness in people as a route to confer immunity. Some diseases have been wiped out completely in the wild by careful monitoring and vaccination. Smallpox, for example, now only exists in a small number of laboratories around the world, and is used exclusively for scientific research purposes.

Another reason we have not been wiped out by wild bacteria and viruses is drugs. Scottish biologist Alexander Fleming discovered the first antibiotic in the 1920s when he saw that a mould that had accidentally grown on a Petri dish seemed to have anti-bacterial properties. He discovered that the mould was producing a chemical, eventually named penicillin, which was widely used to save the lives of soldiers in the Second World War. Drugs to treat viruses took longer to develop. Until the first experimental antivirals to treat herpes came along in the 1960s, there was little that could be done for anyone who developed a viral infection. Vaccinations had reduced the chances of contracting a virus but, once infected, medics could only treat the symptoms while the virus ran its course. In the past few decades, however, a better understanding of virus structure and genetics has given scientists targets to design the drugs needed to attack them, preventing them from replicating in a host.

The microbes fight back

For every immune response that disables the effects of a virus and every drug that wipes out a species of bacteria, another bug will evolve to become resistant to the attack. This evolutionary battle between pathogens and the living creatures they attack has been going on for millions of years, refining the genomes of each side as time goes on. As bacteria multiply, random mutations can make some of them invulnerable to certain antibiotics. Once one bug is protected, the resistance can quickly spread through a population and eventually render certain drugs useless. Our over-use of antibiotics hasn't helped, leading to an increased number of resistant bugs. The worry is that, one day soon, a superbug will emerge that is resistant to all antibiotics.

For our lone terrorist to make a world-killing plague, he would need to find a superbug that had never been seen in the wild before, so that no one would have natural immunity. By cataloguing the things that make individual bugs resistant to different drugs, he could also make sure his plague was resistant to all medications, meaning that it could not be controlled. Then he would step on the plane, and watch as the disease spreads all by itself.

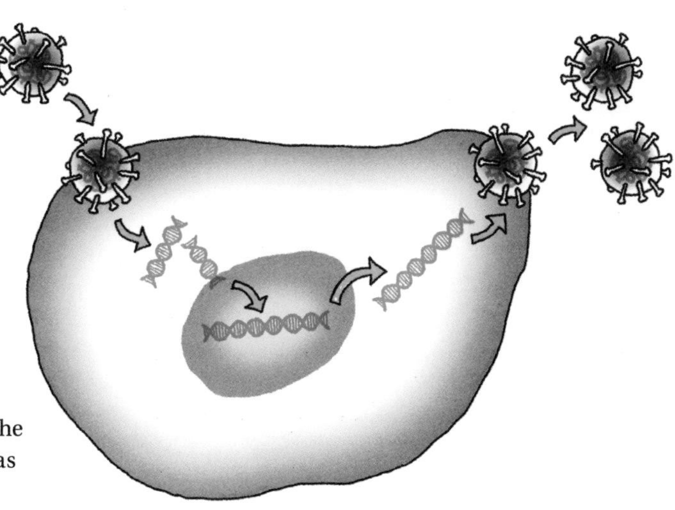

A virus infects a cell, gets copied and these copies burst out. The cell dies in the process.

3 How to live forever

- Why life can kill
- Wasting away
- Programmed to die?
- Reversing the inevitable
- Some tips for a longer life

The German philosopher Martin Heidegger was not far wrong about the inevitability of death when he wrote: 'As soon as man comes to life, he is at once old enough to die.' But how does a living thing die? If it isn't, as the ancients thought, the will of the gods or because some force of vitality is extinguished in a body of earthly matter, can death be defined precisely in terms of biology? What does it mean, at the level of hormones, cells and molecules, for something to age and, eventually, shuffle off this mortal coil altogether? And what can we do to stop it?

Why life can kill

Life is a drag on the body. All those years of eating poisons, fighting illness, getting stressed, breaking bones, sunbathing, refusing to eat vegetables and countless other things that build into a typical life, all take their individual toll on the body. Human cells of all types are remarkable at fixing themselves on a minute-by-minute basis. Whenever there is danger, an array of internal machinery leaps into action, ready to destroy invaders, knit bones together, plug breaks in the skin or repair the DNA inside the cell nucleus.

But our already overworked cells cannot possibly fix everything. One biological definition of death is simply the final result of this never-ending attrition: as something grows older, it accrues more faults and its repair machinery simply cannot fix them all. Perhaps some damage to a piece of DNA leads to a fatal cancer. Or perhaps a perfect storm of smaller faults, each manageable or innocuous on its own, combine to make the body susceptible to a pathogen at a particular moment. If the protective parts of the body cannot work together fast enough, death is inevitable.

Wasting away

The physical state of our bodies over time – from bones, muscles and hearts to brains and immune systems – depends on everything from genetics to the type of environment in which we have chosen to live out our lives. Access to medication is also important, and even a person's level of education has been found to influence life span. But age is its own risk factor for death. There's little doubt that getting old is the biggest single risk factor in contracting life-shortening diseases, from dementia to cancer.

Beyond the age of 30, the human body begins a process of streamlining. This might fly counter to many people's individual experiences of getting older (how many people in their mid-thirties can say they have the same figure as a decade earlier?), but the facts are clear: between 30 and 80, a person will

'Thus all the days of Methuselah were nine hundred and sixty-nine years; and he died.'

King James Bible

lose 40 percent of their muscle mass, and the fibres left behind are somewhat weaker than their youthful versions. It is a similar story with our bones. The strength and mass of the skeleton will rise until the early 30s, after which men lose around 1 percent of their bone mass every decade. This figure is the same for women but, around the menopause, their bone loss speeds up to around 1 percent per year. This alarming bone loss does slow down again to the same rate as men after a few years but the effects are chilling: in five years, a post-menopausal woman's skeleton can age by 50 years compared to a man's of a similar age. Weaker bones are more likely to break. Weaker muscles mean an inability to react appropriately to prevent a fall or jump out of the way of a moving car or bike. Both can have devastating effects because, as the body grows older, it takes ever longer and more effort to effect the necessary repairs.

Cancer affects all age groups but the absolute rate of death goes up with age. In the UK, more than 140,000 people over the age of 70 are diagnosed with cancer every year and more than 100,000 of them will die from it. The most common cancers in this age group are lung, prostate, breast and colorectal cancers. With a rapidly ageing population in all parts of the world, these numbers are only set to increase.

And let's not forget the brain. After the age of 40, this organ decreases in volume and weight by 5 percent every decade. Some people are relatively unaffected by this change, while others might become more forgetful over time and develop neurodegenerative diseases, such as Alzheimer's. If the genetics and environment conspire, a person could end up with conditions such as Parkinson's or Huntingdon's disease.

Programmed to die?

Body cells are dividing all the time. This is obvious when babies turn into children and children grow into adults. But an adult's body cells also refresh themselves on a regular basis, partly to replace the cells that die because of the damage they undergo as we go about our daily lives. One of the most damaging things for a body cell is something that it creates itself: a free radical. This is a highly reactive molecule, a by-product of the metabolic reactions carried out inside cells that turn food into usable energy. Free radicals tear through the body, damaging anything they come into contact with – from the proteins that make up the structures and enzymes, to the fats that surround cells, and even the DNA inside the nucleus. Damage to proteins can cause different symptoms depending on where the free radicals strike. In the kidney they can lead to renal failure, they can cause stiffness in blood vessel walls, while DNA damage results in a cell not being able to produce the proteins it needs to work properly.

As cells carry out their functions, whether they make up the blood, liver, skin or muscle, they will eventually pick up damage from free radicals, poisons or for other physical reasons, and become less efficient at what they do. Many will die. New cells with new machinery are always required to make up for the unhealthy or dying cells. In a process called mitosis, a healthy body cell will split into two over a period of days, each cell an exact copy of the original, now able to do double the work of the original.

But cell divisions also have their problems. Cells that have divided many times will accumulate mutations in their DNA, since no copying process in such a complex molecule can ever be perfect. Most mutations have no effect on the function of a cell but, every so often, DNA damage can lead to uncontrolled cell division and cancer. Cells therefore have several inbuilt limits to how many times they can divide. One mechanism, called apoptosis, is activated when a cell is too damaged to repair for whatever reason. This programmed cell suicide means that it is removed from the body before it can do any further harm.

The mechanism that prevents cells accumulating too many dangerous mutations in its DNA involves the cap on the end of the chromosomes inside cells, called a telomere. Every time a cell divides, the DNA is copied but the telomeres, which are glued to resulting chromosomes a bit like the caps on shoelaces, get shorter. When the cap is too short, the cell can no longer divide. This successive shortening implies a limit to the

Telomeres (the blue caps on the ends of these chromosomes) prevent a normal body cell from dividing too many times.

number of times a cell can divide and, perhaps, an upper limit to the age of a cell. If, after its assigned number of divisions, a body cell cannot divide and replace itself in the event of damage, it will eventually die or continue to work well below its best.

Multiply up by several million cells like this and you can see that the body might suffer. Whether the telomeres are the ticking clock on our life spans has not been completely determined – experiments on nematode worms genetically modified to have longer telomeres showed that they did have longer life spans, but whether the link between telomeres and cell ageing is causal or coincidental is still up for debate.

Reversing the inevitable

Death might be inevitable but the path need not be quick nor painful, however many blocks life might throw our way. Modern medicine has already done a remarkable job of extending our life spans, and the benefits keep coming: by this time tomorrow, your life span will have increased by almost five hours. At the turn of the 20th century, anyone reaching the age of 60 was considered to be near death's door. A hundred years later, they are barely old enough for retirement.

In addition to the scores of treatments already available for diseases such as cancer, hypertension and diabetes, scientists are also working on a range of drugs to deal with the stuff that wastes away. There are already medicines that can prevent muscles and bones wasting away so quickly, and research teams are working on ways to stimulate safely the growth of these vital parts of the body so that older people can lead healthier lives.

Stem cells, the body's master cells that can grow into any type of tissue in the body, also hold promise. Damage caused to organs through accident or disease might one day be repaired by tailor-made cells for a patient, but this is several decades away. Even longer term are potential cures for brain diseases – there are no known cures for dementia and leads are thin on the ground, but who knows what the next generation of brain scanners and neurological drugs will bring?

Some tips for a longer life

All of these treatments, however, deal with the symptoms, rather than causes, of ageing. Can we do anything to slow down or stop the steady march to death? The research here is patchy and inconclusive. Genetic studies show that there seems to be no instruction in our DNA that tells us when to die.

However, there are several genes that are responsible for very different bodily functions that have a cumulative effect to make us age. Here are some tantalizing clues. Restricting how much you eat might help, it seems. In experiments, calorie-restricted rats were found to be physiologically younger, contracted diseases later in life and had an increased life span of up to 30 percent. It is thought that cutting calories switched the rats into a stasis mode of some kind, where growth and ageing are put on hold temporarily.

Experiments in yeast also point to interesting genetic clues for living longer – scientists made yeast cells live six times longer than normal by blocking the action of two genes, one of which controls the yeast's ability to convert food into energy while the other plays a role in directing energy into growing and reproducing. So far, at least ten genes have been discovered in yeast that seem to have some sort of effect on how it ages. Nevertheless, it probably doesn't have to be spelled out that humans are more complex than yeast.

Other (more left-field) ideas to extend human life span to 1,000 years or longer include gene therapy to repair damage, and implanting people with bacteria to clean up the waste and free radicals that build up inside cells as they go about their daily business.

But delaying death is not just a case of high-concept technology that may come in several decades. Achieving a longer, healthy life is all about some simpler actions, such as the quality of the maintenance. Start with a high-quality body (and that means eating your greens, not smoking and doing lots of exercise in your younger days), and there is no reason you can't keep yourself going, if not yet for ever, until well past 100.

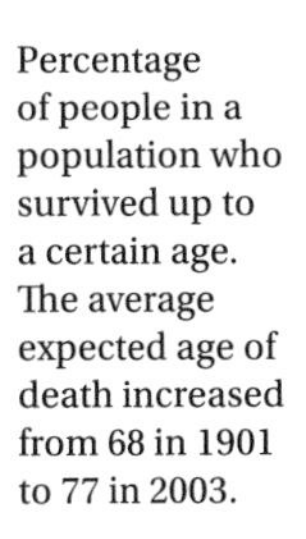

Percentage of people in a population who survived up to a certain age. The average expected age of death increased from 68 in 1901 to 77 in 2003.

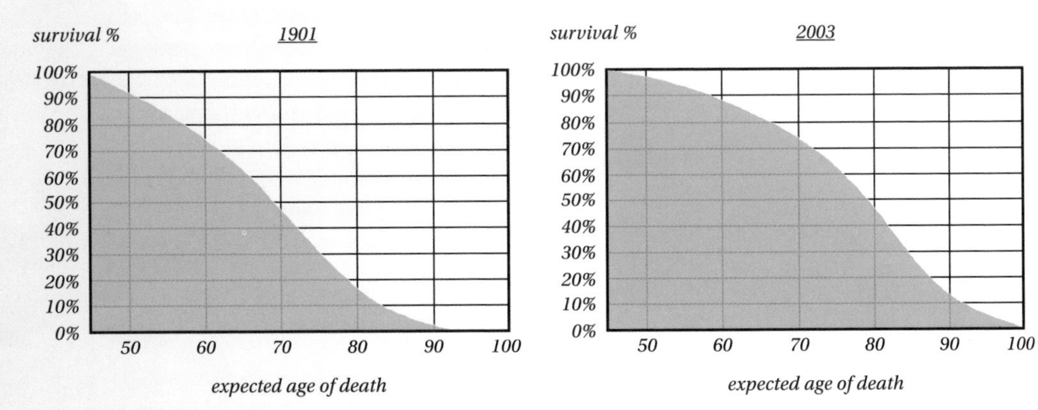

4 How to heal the sick

- What is a drug?
- Molecular action
- Problems messing with the body
- How about treating yourself with no drug at all?
- A drug made just for you?

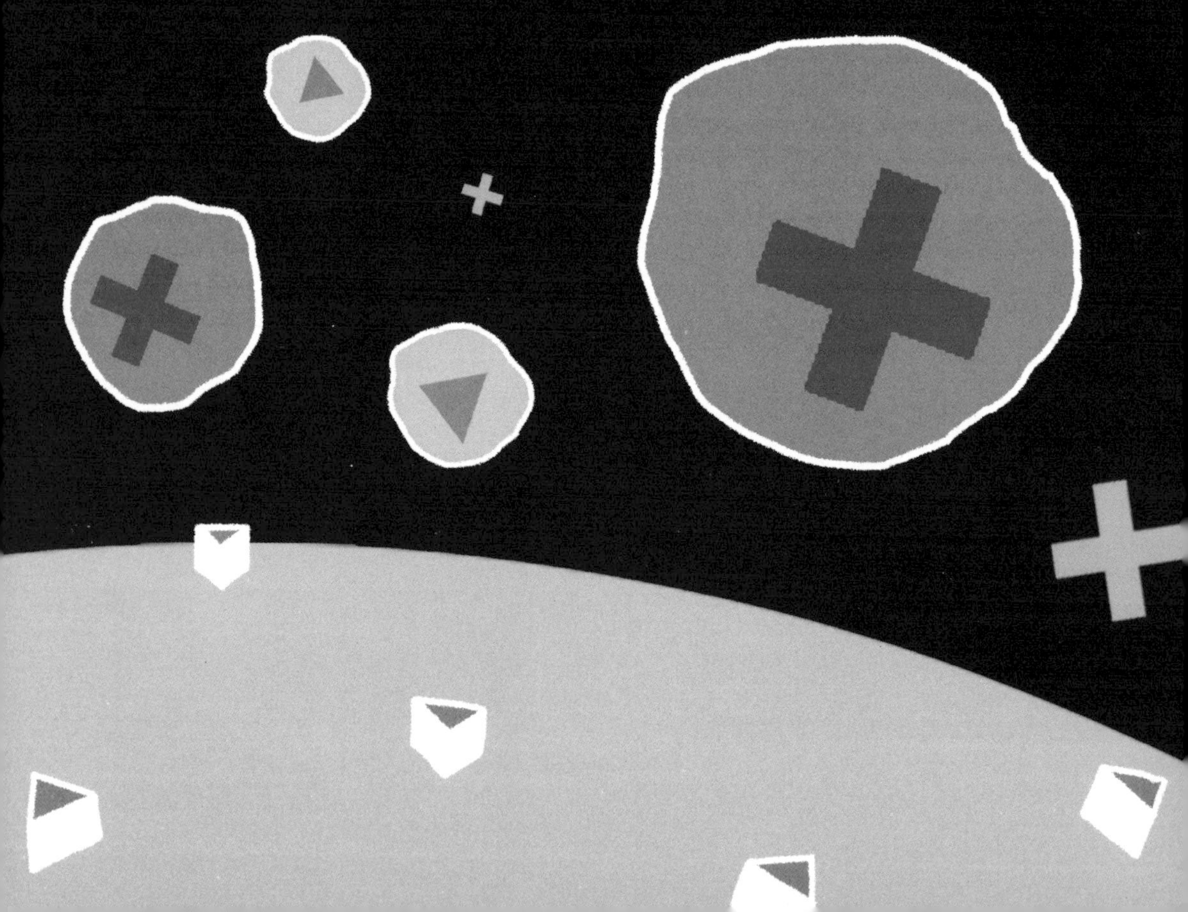

Popping an aspirin at the first sign of a headache is almost an instinct for some. You know how much to take, what to take it for, and you expect it to work quickly and without ill effects, but its origin and make-up is probably the last thing on your mind. The father of medicine, Hippocrates of Cos, thought similarly about the powder he made from the willow tree in 450 BC, but the process of developing drugs has now been refined into a robust, billion-dollar industry that claims to cure everything from erectile dysfunction to cancer.

What is a drug?

The active ingredient in aspirin is a modified version of the chemical salicin called acetylsalicylic acid. It was developed into a marketable drug by the American firm Bayer at the turn of the 20th century and found massive popularity after the Spanish influenza epidemic of 1918. Despite the widespread introduction of other types of pain medication in the past century, such as paracetamol and ibuprofen, aspirin remains popular today partly because of its number of other uses. It has anti-clotting properties in the blood, making it useful in preventing heart attacks and strokes, and some experimental uses include protecting against liver damage and warding off death from cancers, such as those of the breast and colon.

Thousands of medicinal compounds were known to ancient physicians, but central to the study of modern pharmacology is the task of working out how much of a specific chemical is safe to administer to a person and how much is required, over how much time, to deal with a specific problem. Anything available at your local pharmacy has had a long journey from the biochemist's lab. Given their ability to cause significant health effects, drugs are regulated by government agencies and only approved for sale after extensive trials to ensure their safety and efficacy. It takes billions of pounds and many decades to bring a typical drug to market.

It wasn't always like this. Before governments got involved in regulating drugs and checking their safety, people regularly used powerful drugs, such as digitalis, nitroglycerine and quinine for heart disease or insulin for diabetes. Antibiotics became widespread during the Second World War and, by the 1950s, anti-psychotic drugs were in common use.

Today, anything sold as a drug has to undergo a system of approvals. Regulations in most countries require that any candidate small-molecule drug is tested for safety and efficacy in several stages of clinical trials. The early stages involve testing for toxicity and safety by administering it to animals. This is followed by stages of trials in people with and without the condition that the drug is meant to tackle. This part could continue for several years depending on the condition and treatments that the new drug is being compared to.

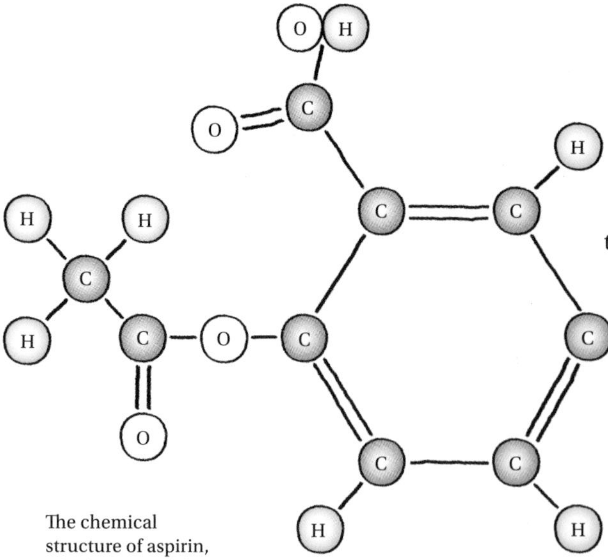

This part can also be conducted double blind, so that no-one knows who is given the new drug and who takes the control treatment. The clinical trials allow researchers to collect data on how effective the new drug is and whether there are any unexpected dangers associated with it that could not have been predicted in lab tests or in animal experiments.

The chemical structure of aspirin, which consists of hydrogen, carbon and oxygen.

Molecular action

A pharmaceutical is any chemical introduced into the body that can interfere with the normal biological functions of the body. Every day, our bodies fizz with chemical reactions, each one meant to carry out a specific step in the process of metabolism. Some of these reactions release the energy required to open and close receptors, which are protein molecules embedded onto the outside of cell walls. Receptors allow specific signalling chemicals, such as hormones, neurotransmitters or a small protein, to latch on board.

When a molecule is bound to a receptor, a bit like a key fitting itself into a lock, it will initiate a response from the cell. On a brain cell it might trigger the release of dopamine, a hormone crucial in regulating everything from the feeling of pleasure to creating memories and fine control of the muscles. A cell will have many types of receptor, each tuned to a specific type of chemical signal. Receptors are also where drugs latch onto the cells, and they do it for the same reasons as any other signalling molecule – to produce a response. A drug might be chosen for a specific job because of its similarity to a specific signalling molecule (it might be a similar shape to the molecular 'key' needed to fit into a specific receptor 'lock'), allowing it to activate a response.

Drugs for high blood pressure typically affect the steps associated with causing the disease, such as heart output and how readily the blood vessels expand and contract. Cholesterol medications target the metabolism and creation of cholesterol. Diabetes drugs look to improve the sensitivity of muscle and fat to the action of insulin, while better-regulating the release of the hormone itself from the pancreas.

Problems messing with the body

Even with safety checks, modern drugs can have copious side-effects. They are, after all, powerful chemicals and anything meant to treat a problem in one part of the body will not stay just in that part. This can lead to unwanted effects such as nausea, headaches, fever or more severe allergic reactions. Estimates in the UK suggest that between 5–10 percent of all hospital admissions are due to adverse reactions to drugs.

Sometimes drugs can even get through clinical trials, become approved and still prove problematic. The pain medication Vioxx, made by drug company Merck, was approved for use for treating arthritis and other pains and around 80 million people are thought to have been prescribed this medication or its equivalent in countries all over the world. In 2004, however, Merck withdrew the drug from the market after concerns that long-term use might increase the risk of heart attack or stroke.

How about treating yourself with no drug at all?

With so many potential problems associated with powerful chemicals, it is also worth considering the therapeutic effects that thinking can have. A placebo is a sham treatment given by doctors that can have measurable effects on a person's wellbeing. Far from being the preserve of charlatans who don't know what they are doing, the placebo effect is a much-studied phenomenon and established in medical practice. Placebos can take the form of normal pharmaceuticals, either pills or injections, but there is no active ingredient inside. In the classic treatment, a patient is given a placebo

Molecules of drugs or natural body chemicals (grey blobs) fit into specific receptor sites on the surface of a body cells (in blue) and trigger specific actions by the cell.

pill and told that it will improve their condition. They're not told that the 'drugs' they have been prescribed are, in fact, nothing more than sugar pills. If the patient ends up feeling better after the treatment, it could be due to a subjective feeling that, since he has taken a treatment, it must have worked.

The placebo effect is not perfect. It involves deception by the doctor, which raises valid ethical questions about whether the patient's best interests are being taken into account. And whether placebo has a real physiological effect is hard to pin down – several scientific reviews have been inconclusive. Still, experiment after experiment shows that placebo can sometimes be as powerful as pharmaceuticals, but with the advantage of few or no side-effects. A placebo pill given as a stimulant will raise heart rhythm and blood pressure but, presented as a relaxant, has the opposite physiological effect. Alcohol placebos can make you feel drunk and lose coordination. Appearance can also make a difference: blue pills are better relaxants while red ones work as stimulants; big pills can increase the size of the placebo effect; pills seem to work more often than tablets, and injections seem more effective than pills.

A drug made just for you?

Making and marketing drugs is big business, and many of the ideas have, until now, come from observations of causal mechanisms of a disease, or of how microbes affect the body. As such, they are not aimed at specific individuals and can have wildly different effects in different people. This can be no more problematic than two people requiring different doses of a drug for it to have the same effect but, in extreme cases, the effects of a single drug could be to treat one person while causing a severe allergic reaction in another.

In an attempt to understand these differences, scientists have turned to the human genome – the sequences of DNA that code our individual bodies and which greatly influence how our individual metabolisms deal with different drugs.

Human genetic variation is vast at the level of individual mutations, and there are versions of genes for everything from hair colour to the regulation of metabolism. These variations are small differences in the genetic code between people and are not usually indicators of illness or abnormality. But many will have minute physiological manifestations, in how efficiently particular receptors or

'Pharmacogenomics, which from my perspective has been one of the most promising areas of personalized medicine, has also turned out to be extremely complicated, not that we shouldn't have known that.'

Francis Collins

parts of the metabolic chain work, for example. In a sense, doctors have been doing this for a while. Family histories can tell them about diseases you might be at a higher risk of developing. And they have always given us advice on eating healthily and exercising regularly to avoid the ravages of obesity.

Two factors have made genetics more relevant to medicine in recent years: scientific journals are continually publishing new and better knowledge of the links between genetic variations and common disorders; and the cost of DNA sequencing is dropping quickly. Together, this increasing knowledge heralds an era when doctors will know exactly which variants a patient has, and will allow them to use this knowledge – along with the usual factors such as age, weight and allergies – to prescribe a specific drug that has a much higher chance of working.

Sorting through the masses of information coming out of genetic analyses, though, is going to take some time. Some diseases are caused by a single DNA mutation, but there are not many of these. Most diseases are a complex interaction between scores of genes and lifestyle factors. Early results in mapping variants have been promising, already proving useful in deciding on doses for drugs such as the anti-coagulant warfarin. Other examples include variants in the genes involved in making and processing cholesterol, which make some people less susceptible to the effects of drugs such as statins, in which case a doctor might prescribe them something more suitable.

The millions of chemical reactions at work in our bodies perform a remarkably choreographed dance to keep us functioning, a complex biochemistry that creates every wrinkle of the thing we call life. Very occasionally, one or more of these things may go wrong or may not work properly. Fortunately, we are finding ever more intricate ways to locate and fix these errors, to manage the problems and to tweak our biochemistry to make us feel better.

5 How to build a brain

- Think of an apple…
- Our impoverished perception of the world
- Sound and melodies
- Learning and memory
- Topographies
- Building blocks
- Frontiers

The human brain is an unimpressive-looking mass of fatty tissue and water, weighing just over a kilogram. But this mushy organ is what sets us apart from the rest of the world's animals, the thing that we can rightly say is at the apex of the process of biological evolution. This bundle of fibres and connections buzzing with electricity has given our species the ability to write **A Midsummer Night's Dream,** ***win gold medals at the Olympics, discover quantum physics and walk on the Moon.***

Think of an apple...

Very simple things can be caused by some complex physiological actions in the brain. A thought of an apple, say, will immediately bring to mind a round object that fits easily into a hand. It might be red or green or a combination (perhaps with some russet too), and you might wince a little as you imagine the sharp acidity of biting into it.

It's instructive to ask how you came to think of all those things, how you managed to bring up the memories and learning associated with the apple. Perhaps there is an 'apple neuron', which stores everything you need to know and which fires whenever you see (or think) about the fruit? This implies that there must be collections of neurons storing the relevant information for each object in each of our memories. There would be a neuron for your favourite song, one for each of your friends, one for your mum.

But there's a problem with that idea. An apple neuron would have to 'recognize' a whole slew of properties including size, shape, taste and smell, as well as being able to fit that data into the overall concept you might have of fruit, food and hunger. Never mind whatever memories you might have of apples or any disassociation from the Apple computer company or Apple Records. And how does it figure in any thoughts you have of your mother's apple crumble? Do all the various individual neurons for apple, mother and crumble fire at once? Or is there another, separate, neuron for 'mum's apple crumble'? So how did custard get into that thought?

It's a mess. Instead, neuroscientists think that all information (from knowledge of objects or smells to learned skills and those fond memories of childhood summer holidays) is stored in networks of neurons. More specifically, in the connections between synapses. New connections are made and broken every day depending on the experiences we come across and the practice we do. The more often we use a synapse, the stronger it gets.

Neglected synapses get deleted. Networks in different parts of the brain, each responsible for a different aspect of perception or sense, work in parallel at every moment to create the picture of the world around us.

Our impoverished perception of the world

When light hits your eyes, it passes through your cornea, is focused onto your retina by the lens and then processed by a specialized cell called a photoreceptor. Some of the 125 million of these that line the back of the eye absorb the light and send electrical signals to the nearby neurons. The signals pass along the optic nerve and into the visual cortex in the brain. The left half of the image in each eye registers in the right hemisphere of the brain and the right half of each image is processed in the left hemisphere.

How visual information is processed in the brain is not fully understood but scientists know that it requires a lot of work – around a quarter of the brain's efforts are engaged in dealing with the input from the eyes. And yet we can be easily misled. Magic tricks, for example, throw our carefully constructed mental picture of the world into disarray. Objects seem to float in mid-air and coins and cards vanish in front of our eyes, because our brains are selective about which bits of sensory information to process.

Images from the right-hand side of our visual field are processed by the left-hand side of our brains and vice versa.

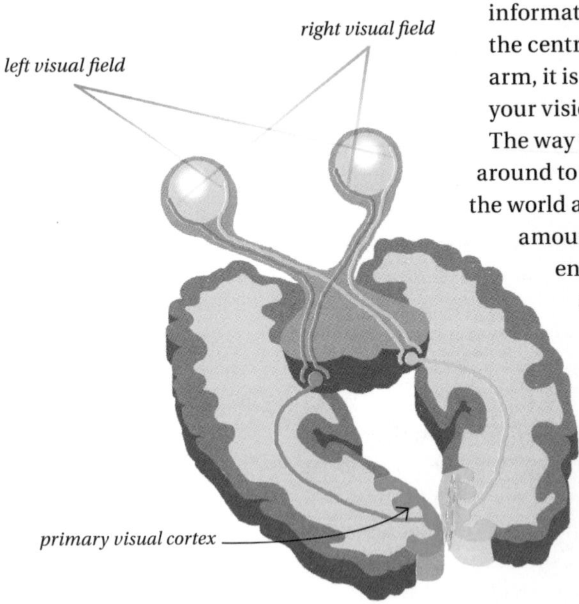

right visual field

left visual field

primary visual cortex

Scientists know that we only receive high-quality information from the area we are fixated on, right in the centre of our field of view. If you stretch out your arm, it is about two thumbs' width at the centre of your vision. Everything else is pretty much blurred. The way we compensate for this is to move our eyes around to fill in the gaps and create a better picture of the world around us. Our brains filter out a huge amount of the sensory input flooding in from our environment. You could be looking at something without being aware of it, if your attention is focused elsewhere.

Sound and melodies

While sight is, in some sense, impoverished and expensive to process, sound is all about careful processing. The parts of the brain dealing with this sense contain millions of neurons that recognize different types of sound – some respond to pure tones, others to complex musical notes. Some neurons fire when we hear rising frequencies, others when the sound is short rather than long. Yet other neurons combine the information processed by the other parts of the brain in order for us to recognize a word or sound. Though sounds are processed on both sides of the brain, scientists have found that the left hemisphere tends to become specialized for understanding and producing speech. Damage to the left part of the auditory cortex, therefore, can leave someone able to hear but unable to understand language.

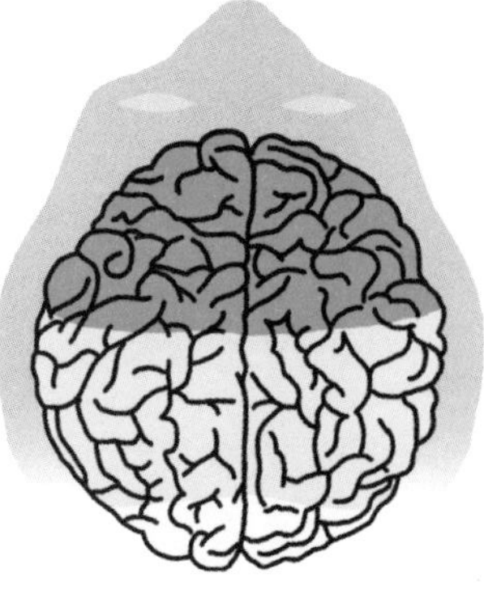

Learning and memory

Memories are the brain's way of storing information, while learning is the biochemical process by which memories are laid down or changed. The ability to learn and recall everyday facts and events is called declarative memory, which is controlled in the cerebral cortex. The information gets to this part via the prefrontal cortex, which holds temporary information and is also active when we think about old memories. Memories of events and personal experiences, called episodic memory, are kept in networks of neurons in the medial temporal lobe and different parts keep track of the 'what, where and when' information in any remembered event.

The cortex is made of four lobes: (clockwise from dark blue part, above) frontal, parietal, occipital and temporal lobes. The grey area is the cerebellum.

The formation of memory involves strengthening the synapses between networks of neurons. Scientists at the University of California, Los Angeles, have even watched these networks being made, albeit only in the relatively simple brains of sea slugs. No-one has yet seen a human make a memory but, with ways of looking at the brain in action always improving, that observation cannot be long in coming.

Topographies

The outermost section of the brain, the cerebral cortex, consists of four parts: the occipital, temporal, parietal and frontal lobes. These regions control many of the higher sense functions, such as hearing, vision and speech. The internal structure of the brain is more varied. The forebrain is given over to the cognitive tasks we associate with higher intellectual abilities, including thinking, planning and solving problems. At the very centre of the brain is the thalamus, a clearing house that coordinates all the information coming into the brain from the furthest reaches of the body. Slightly ahead of that is the hypothalamus, the switchboard for regulating internal systems, which it does using information fed by the autonomic

'Belief has been a most powerful component of human nature that has somewhat been neglected. But it has been capitalized on by marketing agents, politics and religion for the best part of two millennia.'

PETER HALLIGAN

nervous system. It then responds with instructions by sending nerve impulses back, or instructing the pituitary gland to release hormones. Underneath that is the hippocampus, which has a role in memory, and the amygdala, one of the most primitive areas of the brain, which has a role in emotion and warning us of dangers in our environment. At the back of the brain, just above the meeting with the spinal cord, is the pons and medulla oblongata, which helps control respiration and heart rhythms. The cerebellum is also here, helping to control movement and cognitive processes that might require very precise timing.

The brain communicates with the rest of the body through the nervous system, a network of cells similar to brain cells, which extends gossamer-like to the tips of the fingers and the ends of our toes. Think of this as the extension of the brain into the body. Using electrical signals, the brain gathers information from all extremities and all organs, processes a response, and sends instructions back. Voluntary body movements and senses such as touch and pain are controlled by the peripheral nervous system, which attaches to the brain via connections along the spinal column. The central nervous system (brain and spinal cord) is also connected to the organs via the autonomic nervous system.

Building blocks

The basic unit of the brain, and vast reaches of the nervous system, is the neuron, or nerve cell. Everything we think, feel and remember, every conscious and unconscious action and every movement we make is, at its basic level, down to the myriad interconnections between groups of neurons somewhere in the body. Neurons are much like any other cell in the body (with a nucleus and mitochondria for metabolism), but where they come into their own is with their axons and dendrites, which can extend for up to 1 m (3 ft) through the body, connecting neurons across long distances from the spinal cord, say, to the farthest toe. Axons are the electrical conductors extending from the neuron's cell body, dividing in several directions before ending in a nerve terminal. On the end of a nerve terminal is either another neuron's cell body or else a dendrite, a tendril of a different neuron somewhere further away. These sites of connection are called synapses. Neurons communicate by firing electrical signals at hundreds of kilometres per hour along their axons, for distances that can range from a few millimetres to a metre. When the signal reaches a nerve terminal, it triggers the release of neurotransmitters, the nervous system's chemical messengers. The chemical molecules diffuse across the space between synapses and attach themselves to receptors on the surface of the target cell – this could be another neuron but also a muscle, gland or organ cell.

All body cells are covered in a multitude of different receptors, each with a distinct shape and each activated only by the chemical messenger that fits their shape. The receptors act as the gatekeepers to a cell and, when activated, instruct the cell to carry out a specific action. If the target is a neuron, its receptor might tell the cell to pass on the electrical signal it has just received; if it is a muscle, the activated receptor might cause a muscle unit to contract; if it is an organ, it might be instructed to kick start a chemical reaction.

The chemical circuits that connect neurons are key to how the brain stores information, how behaviour and impulse work and in understanding the biological basis of brain illnesses. Deficits in the neurotransmitter dopamine, for example, can lead to Parkinson's disease where people experience muscle tremors, rigidity and have difficulty moving. Low levels of serotonin have been linked to depression.

Neurons can be up to 1 m (3 ft) long and connect to many other cells or organs.

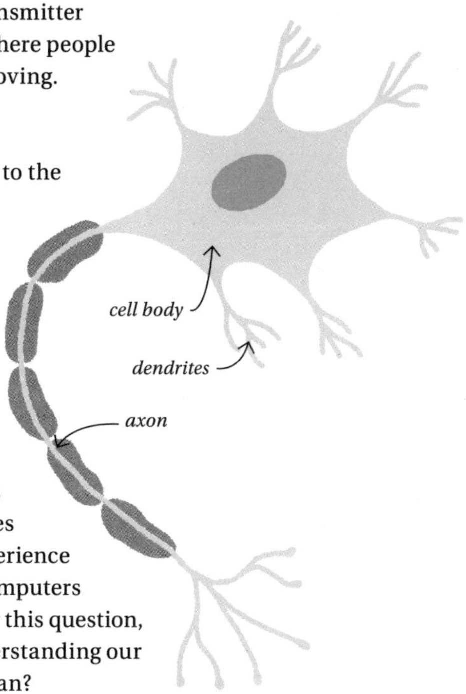

Hormones are another way that the brain sends messages to the disparate bits of the body. These chemicals are the equivalent in the endocrine system of neurotransmitters in the nervous system. The brain has receptors for all the major types of hormones and uses these chemicals to regulate some of the basic behavioural functions in the body such as sex, emotion, response to stress, growth, reproduction and metabolism.

Frontiers

Discovering the physical structure of the brain, which has taken decades of work already, is barely the start. How does the brain turn the drone of electrical activity into our experience of the world? Better brain scanners and more powerful computers that can model networks of neurons are starting to answer this question, taking neuroscience towards the next crucial step in understanding our brains – how does this mass of wetware make us feel human?

6 How to turn sunbeams into oak trees

- Leaves, roots and stems
- Acorn to tree: a timeline
- Living off the Sun
- Guardian of the centuries

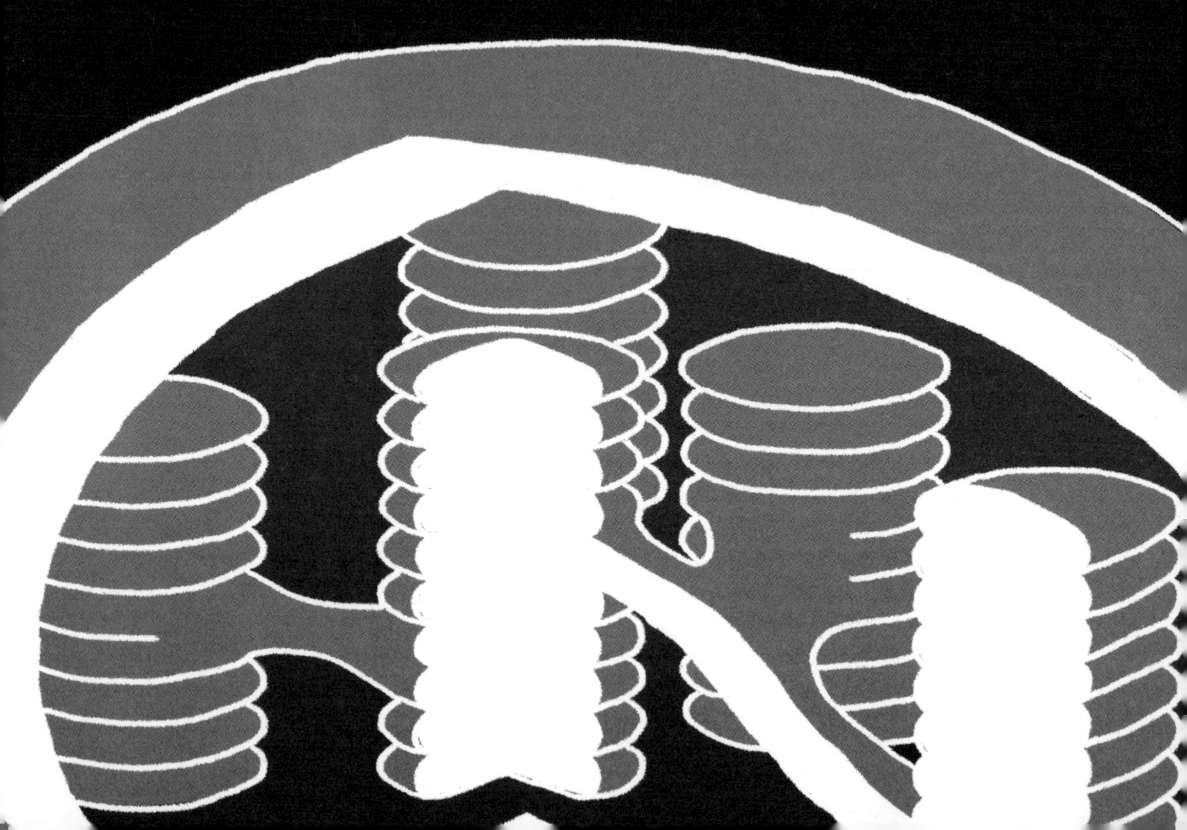

Trees are a testament to successful biological engineering through evolution. Swaying in a breeze, with thick, leaf-encrusted crowns dappling sunlight onto the ground, they are symbols of calm, strength and patience. But underneath that exterior is a factory of activity: cells are dividing, leaves buzz with electrical charge, sophisticated hydraulic systems move huge volumes of water against the force of gravity every day, and chemical reactions vital to all life on Earth are being carried out every second.

Leaves, roots and stems

Any tree can be divided into three main bits: the roots, the stems and the leaves. Each serves a specific function to keep the tree healthy, fed and growing. The roots, spreading out underground in a network as intricate and extensive as the branches above ground, hunt out water and dissolved nutrient minerals from the soil. A central root comes out directly under the trunk, which then forks endlessly and in all directions into underground branches and stems. The tiniest elements of the root are as thin as hairs. With the thinnest of cell walls, they are grown anew every spring to absorb water and minerals from the soil more easily. By autumn, once the growth phase of the tree is over for the year, the hairs wither away.

The above-ground part of the oak is held up by the woody trunk that is, in effect, the first, biggest and most important of the tree's network of stems. The thickest branches come straight off the trunk with further stems splitting off these further along their own length into ever-smaller versions of themselves. The composition of all these structural elements is shared: a double-layered bark for protection, cambium, sapwood and heartwood.

At the tree's centre is the heartwood (blue), surrounded by sapwood (which includes the growth ring), then the phloem and finally the external bark.

The bark protects the innards of the trunk and branches from any damage that could be caused by objects or animals that encounter the tree. It also prevents the tree from drying out. The familiar outer layer, which can be split and rough, is made of dead cells, much like the epidermis in our own skin. But the inner bark is very much alive, carrying food and water to the rest of the tree. This part, the phloem, is the tree's circulatory system, the part that glistens if you chip off a piece of the outer bark of a tree. In the cambium, the main stem of the tree does its growing, in tune with the shifting seasons. Both bark cells and wood cells are later produced in this, during periods of growth that alternate with periods of dormancy. The results of this cycle are the characteristic annual growth rings in trees, and a typical oak tree might gain between 1.5 and 2.5 cm (up to 1 in) to its circumference every year, depending on how much water and nutrients have been available that season. These rings do much more than just identify how old a tree is or how much it grew in a certain year, however. Scientists have used tree-ring data to track the changes in carbon dioxide in the atmosphere throughout history, and to date major environmental events such as droughts, floods or volcanic eruptions. Tree rings only exist in trees that grow in changing seasons – in the tropics where seasons are more-or-less the same all year round and there is near-constant sunshine and rainfall, trees might not have growth rings at all.

Under the cambium is a layer of cells called the xylem, the woody part of the tree that is split into sapwood and heartwood. As its name suggests, the first of these carries sap, a mixture of hormones, minerals and sugars dissolved in water, between the leaves and roots. Sap can be a commodity in its own right, along with the wood, in trees such as birch or maple. More generally, sugary sap is also a nutritious food for many insects and birds. Heartwood provides the tree's internal strength, a consolidation of dead cells that allow the tree to stand upright. It is usually darker than the living sapwood that surrounds it, and is much more resistant to decay than other parts of the tree.

And finally, leaves are the tree's food factories. The oak's familiar lobed, oblong leaves are made from cellulose-encased cells that contain chloroplasts, miniature factories where photosynthesis takes place. This process involves combining carbon dioxide and water (plus sunlight) to produce sugars.

Acorn to tree: a timeline

The oak is a flowering plant, also called an angiosperm, a term that refers to any tree that produces seeds inside fruits or nuts. Trees such as conifers, which produce seeds that are not encased in any way, form the other major class of trees, called gymnosperms. Most conifers tend to have needle or scale-like leaves and are evergreen, keeping their leaves throughout the year. They grow in colder climes and tend to have softer woods than the trees from angiosperm seeds. In contrast, angiosperms are deciduous, growing in more temperate climes, shedding their leaves every year and producing much harder woods.

If the weather has treated it well, a mature oak tree will produce around 50,000 acorns in a year, each starting off in the spring as a catkin, a slim cylindrical cluster of flowers. Acorns contains the seed (and therefore the genetic material) for a new oak tree and also some of the nutrients it needs to make its start in life, bound up in a seed coat for protection.

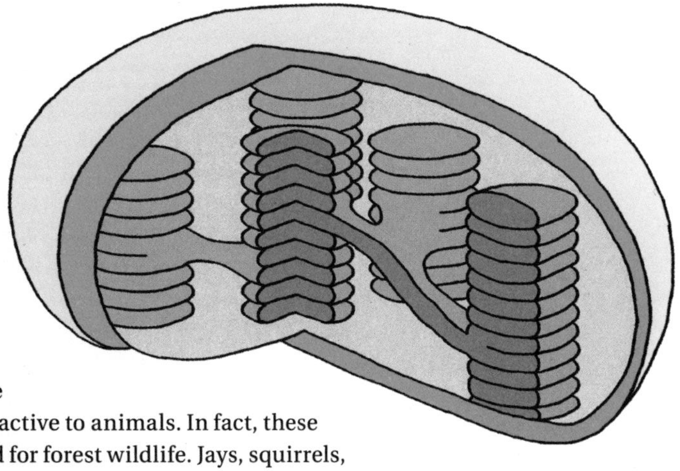

Acorns are rich in carbohydrates, proteins, fats and minerals so that the embryonic plant inside has the best chance of germinating – but the combination of nutrients is also attractive to animals. In fact, these nuts are an important source of food for forest wildlife. Jays, squirrels, pigs, deer, bears, ducks and pigeons all tuck into large amounts of acorns for nourishment.

A chloroplast contains stacks of circular thylakoids, where the reaction to convert sunlight and carbon dioxide into sugars occurs.

If an acorn manages to survive the appetite of the nearby wildlife, it will fall to the ground and, when environmental conditions – including water, oxygen, temperature and light – are just right, the seed inside will germinate. This process starts when the seed takes in lots of water. This causes the seed coat to split, creating an opening through which the various shoots can begin to emerge. Inside the seed is the embryo, an immature plant that consists of a tiny leaf, root and stem. When they are activated during germination, the cells in these embryonic organs begin to multiply, using food from the nut to divide.

As it grows, the embryonic root searches downwards, the start of a major network that will one day anchor the oak tree into the soil. The tiny stem pushes upwards through the gap in the seed coat, breaking through the surface of the earth to find a source of light for its tiny leaves to start photosynthesis. As the plant begins to make its own food using sunlight (rather than relying on the small supply contained in the acorn), its stem will grow longer and small buds will appear along its length. Each one will sprout a new stem that will gradually thicken into a branch, then split again to produce even more stems, branches and leaves. The whole time, the original stem, the one that broke out of the seed, is turning into the trunk, and getting thicker and taller thanks to growth in its cambium.

At the top two-thirds of the oak are its thousands of leaves, which make up the canopy of a mature tree. A mature oak can grow (and eventually lose) around 250,000 leaves per year, and each one provides vital functions such as regulating the circulation systems to move nutrients around the tree.

Water evaporates out of minute holes on the underside of the leaves, called stomata. To replace this lost moisture, fresh water is drawn up through the roots. This not only moves fluids from one place to another – the loss of water from the leaves also prevents the tree from overheating. An oak can soak up around 1,500 l (3,000 pints) of water every day. The leaves also carry out one of the most important processes of all – making all the food a tree needs through photosynthesis.

Living off the Sun

Every part of the growth in the oak tree is fuelled by simple sugars that are made in its leaves. Photosynthesis is one of the most remarkable – not to mention most useful – chains of chemical reactions that has ever evolved for the maintenance of life, and not just for the life of the oak tree in question or even the lives of the myriad other types of plants and algae that can carry out the process by themselves. Virtually all of the energy used to create and sustain every kind of life on our planet depends on photosynthesis at some point in the chain.

Water is absorbed from the ground by root hairs and lost through leaves via transpiration.

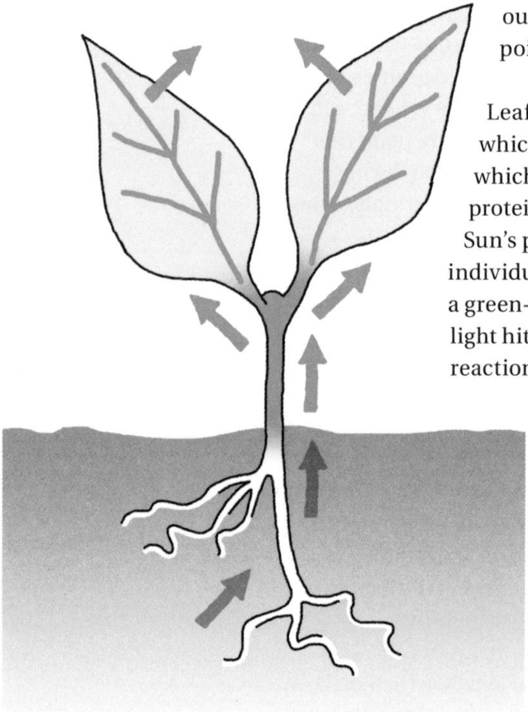

Leaf cells in the oak tree contain chloroplasts, which are mini energy-harvesting factories in which a series of chemical reactions catalysed by proteins spend every daylight hour storing up the Sun's power. There could be anything up to 100 individual chloroplasts in each leaf cell, each containing a green-coloured compound called chlorophyll. When light hits the leaf, it is absorbed by one of the many reaction centres in the chloroplasts, and then stored in one of two ways.

Some of the energy is used to make adenosine triphosphate, a molecule that can be stored by the plant and re-released quickly to power some of its basic metabolic functions. The remaining light energy is used, via chlorophyll, to split water into hydrogen and oxygen, a process that also releases lots of electrons. The oxygen is vented into the air (most of this gas in Earth's atmosphere came via this process in plants), while the hydrogen and electrons are used to convert carbon dioxide, taken from the outside air, into glucose. Once this sugar is dissolved in water, it can be transported via the phloem to whichever parts of the plant need it.

'What drives life is ... a little electric current, kept up by the sunshine.'

ALBERT SZENT-GYÖRGY

Chlorophyll absorbs light waves, mainly at the red end of the electromagnetic spectrum, leaving the rest to bounce off the plant and into our eyes, hence the green colour of most leaves, including oak leaves. Chlorophylls are not the only pigment used by plants to absorb light – carotenes and xanthophylls do similar jobs for other plants.

Photosynthesis is a hugely efficient mechanism for trapping the sun's energy and making it useful for life. Scientists estimate that the world's plants have the capacity to process 100 terawatts of power, many times the requirements of all of human civilization. This process also sequesters more than 100,000,000,000 tonnes of carbon into plants every year.

The oak will use the sugars to power its basic life processes and also to build new bits of itself – joining many hundreds of glucose molecules together produces cellulose and lignin, the two main components of the oak's golden-brown wood.

Guardian of the centuries

Oaks can live and grow for many hundreds of years, turning into majestic plants 40 m (130 ft) tall. Trees in general are the longest-living life forms on Earth. Different species grow at their own rates and will end up at a range of sizes. The Alaskan cedar can live for more than 500 years, reaching a height of 30 m (100 ft), if it is not bothered by disease or parasites (or destroyed in natural disasters, such as fire or droughts) over its lifetime. The redwood trees of north America can live for thousands of years. The biggest tree in the world is a Giant Sequoia in California called General Sherman, which stands 83.7 m (275 ft) high.

Britain's oldest oak is the Bowthorpe Oak in Bourne, Lincolnshire. At more than 1,000 years old, it was just a sapling when William the Conqueror sailed across the English Channel to beat the army of King Harold in 1066. The Angel Oak, a sprawling Southern Live Oak 20 m (65 ft) tall near Charleston in South Carolina, is thought to be more than 1,400 years old – more ancient than any man-made object on that continent. Both these trees have survived centuries of change in the world, standing firm through hurricanes and rains, and as human empires rose and fell. Imagine if trees could speak – what stories they could tell.

7 How to become invisible

- What is light?
- The electromagnetic spectrum
- Unweaving the rainbow
- From one place to another
- On to the invisible

The key to invisibility lies in knowing what light is and manipulating how it behaves around objects. We cannot see the molecules in the air because they naturally do not reflect visible light and we see through glass because most visible light passes right through it. If you can prevent light from reflecting off you and into the eyes of your observer, you are, for all intents and purposes, invisible to that person. And, fortunately for all fans of* Harry Potter *and* Star Trek, *invisibility cloaks are entirely possible to make.

What is light?

The ancient Greeks thought we saw things because of a mysterious substance emitted from our eyes. This stuff would interact with similar stuff coming out of lamps and candles, allowing us to 'see' the source of a light. If a non-luminous object, an apple for example, happened to be in the location where the substances interacted, it would be stimulated to express its colour. By the Middle Ages, the Persian scholar Ibn al-Haytham began to formulate another idea, more in tune with the understanding of light we have today. He suggested that sight was something to do with rays coming into, rather than being emitted from, the eye. He demonstrated that the rays travelled in straight lines, and wrote several important works, including the *Book of Optics*, which would influence Western scientists hundreds of years later.

> *'In so far as a scientific statement speaks about reality, it must be falsifiable; and in so far as it is not falsifiable, it does not speak about reality.'*
>
> Karl Popper

The British natural philosopher and polymath Robert Hooke took up the mantle next, publishing his theory that light was some sort of wave in the 1660s. His contemporary, the Dutch mathematician and physicist, Christiaan Huygens, insisted that the light waves must travel through something, which he hypothesized must permeate all of space, and which he called the 'luminiferous aether'. The idea of an aether persisted until 1887, when US physicists Albert Michelson and Edward Morley set up an experiment to measure the properties of the mysterious substance. Using an interferometer, they combined light waves travelling in different directions, an attempt to see how the properties were different depending on how the waves moved through the aether. Unexpectedly, their work proved that the aether could not exist – light, it seemed, was a wave that did not need a medium through which to propagate. Around the same time, the British physicist Michael Faraday noticed that light rays were affected by magnetic fields. This inspired Scottish physicist James Clerk Maxwell to begin work on a set of mathematical laws to describe electromagnetic (EM) forces, which he used to prove that light was a vibration in the EM field.

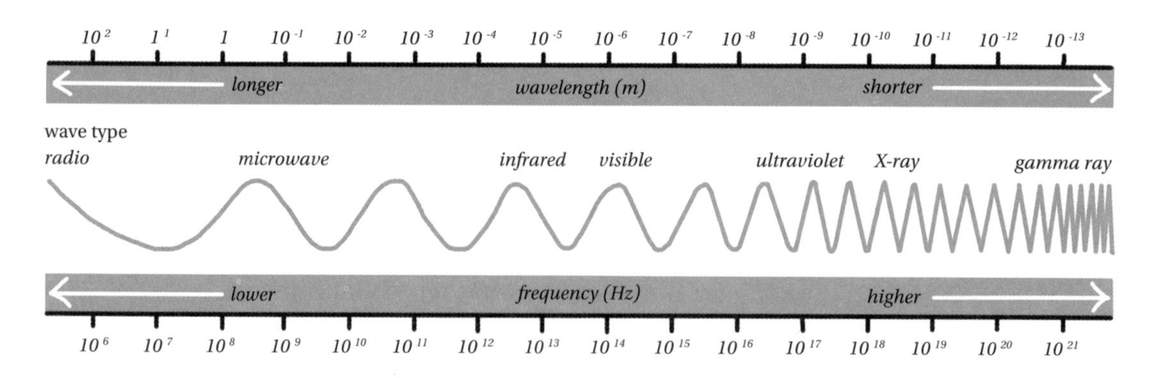

The electromagnetic spectrum ranges in size and energy from radio waves to gamma rays.

The electromagnetic spectrum

We know today that light is an electromagnetic wave that travels at 299,792,458 m/s through a vacuum. Think of the shape of a light wave as identical to water waves on the surface of a lake. Taking a vertical cross-section of those waves, you will see peaks and troughs as the vibration moves through the water. The distance between successive peaks (or troughs) is the wavelength. Visible light, which our eyes can detect, is in a narrow range of wavelengths between 390 and 750 nanometres, and the different wavelengths are interpreted by our brains as different colours. The shortest wavelengths are blue and violet, the longest wavelengths are red.

This range sits in the middle of the EM spectrum, which goes out on either side to several orders of magnitude. At the smallest end are gamma rays, with wavelengths smaller than an atom, while the longest are radio waves that may be several kilometres long. Starting from the longest wavelengths and going down, EM radiation is categorized into: radio waves, microwaves, infrared, visible light, ultraviolet, X-rays and gamma rays. The shorter the wavelength, the more energetic (and dangerous to life) the radiation becomes.

Unweaving the rainbow

We see objects in the world because they reflect light waves. When a light bulb shines onto an object, that object will absorb some of the energy and scatter the remainder in all directions. How much absorption and scattering an object does will determine what that object looks like to us. A mirror or piece of polished glass looks shiny because it does not scatter light much, instead reflecting away the incoming waves at roughly the same angle they fell onto it (we say that the angles of incidence and reflection are the same).

Colour is the result of light absorption. A banana looks yellow because the light it reflects is subtly altered from the original sunlight that hits it. If the banana had reflected all of the sunlight, it would look white, but its markedly yellow hue means the fruit has absorbed the incident light energy at the red, blue and green wavelengths of the visible EM spectrum. If there are black spots, those bits have absorbed virtually all of the incoming light. Scattering can also cause colour. It explains why the sky is blue, for example: the shorter wavelengths of light coming in from the sun (remember, these are the blue and violet colours) are more easily scattered by the moving atoms in Earth's atmosphere while the remaining wavelengths pass right through the air. The human iris and the iridescent feathers of some birds and butterflies get their shimmery colours because they scatter light.

Go back to the analogy of water waves for a moment. What happens when two distinct waves meet on the surface of the lake? When two peaks coincide, they combine to form a bigger peak, and it is the same for a trough. If a peak meets a trough, though, they will cancel out if their amplitudes are the same. This process, called interference, results in the formation of an entirely new wave distinct from the two waves that made it.

From one place to another

When light travels from one medium to another (air, for example, to glass), its speed changes. If a light beam hits the boundary straight on, then no-one would notice the difference other than the light taking slightly longer to get out of the other side. But if the light hits the air-glass boundary at an angle, the light beam will turn through an angle as it passes the boundary, a process called refraction. How much the light bends is governed by a material's refractive index: air has a refractive index of 1, for water it is around 1.3 (which is why a straw looks bent in a glass and fish look closer to the surface of a pond than they actually are). Diamonds have a refractive index of 2.4, hence their unique sparkle. All transparent materials have a refractive index greater than 1.

Lenses also work by refraction. Bending light waves can create the illusion that something on the other side of a polished, carefully shaped piece of glass is bigger or smaller than it really is. Refraction is also the inspiration for Pink Floyd's album cover for *Dark Side of the Moon*. Newton first used a glass prism to split white light into its constituent colours. Different wavelengths of light bend by different amounts at the glass-air boundary, with blue turning through the smallest angle and red light bending the most. Pass white light through a prism and, on the other side, a rainbow of colours will emerge.

Light going from a lower to a higher density material will bend towards a line at right angles to the boundary called the normal.

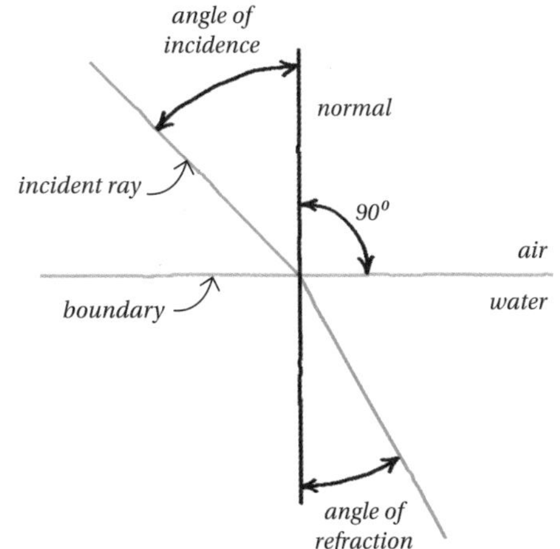

Waves do even more strange things when they reach an aperture that has a similar size to their wavelength. Imagine a beam of light nearing a barrier that is entirely opaque, except for a very narrow slit that is about as wide as the wavelength of the light in the beam. What happens when the light beam reaches the barrier? You might think that most of the beam would be blocked, while the small part unencumbered by the barrier would carry on as before.

But the waves don't do that. Instead, the portion of the wave that goes through the narrow slit spreads out as it emerges on the other side. Parallel light waves encountering a single gap will emerge from the other side as a series of circular waves, like ripples in a pond spreading out from where a stone has been thrown in. Two slits close together in the barrier will result in two sources of circular waves, and the pattern of highs and lows on the other side will be altered as the waves interfere with each other. Each diffraction grating (the name given to the unique pattern of slits in an obstacle) has its characteristic effect on the light waves coming into it. Far from being an annoyance, scientists have put this property of light waves to good use in working out the structure of molecules that are too small to be imaged in any other way (using visible light, for example).

By examining the diffraction pattern of X-rays (which have much smaller wavelengths than visible light) after they have passed through a crystal, a protein molecule or even a DNA molecule, it is possible to work backwards to infer the structure of the obstacle. In these instances, the molecules act as three-dimensional diffraction gratings, the shape of which can be calculated by comparing the light waves on either side. British physicist Rosalind Franklin used X-ray diffraction techniques in the 1950s to create the images of DNA molecules that Francis Crick and James Watson subsequently used to deduce the shape of the double helix.

On to the invisible

So now that we know how light interacts with matter, we can put this knowledge to use in making things invisible. Rudimentary invisibility cloaks work by forcing electromagnetic waves to flow around an object instead of interacting with it in the normal way. When the waves reach an observer on the other side of the object, they arrive unaltered. Thus it is impossible for an observer to 'see' the object.

To do something similar for light waves is possible but, because of the tiny wavelengths involved, scientists have had to develop intricately patterned composite materials that can bend light in unexpected ways. The theoretical basis for these strange 'metamaterials' came from British physicist John Pendry. In the 1990s, he proposed and designed materials that would have a negative refractive index, coming up with the mathematical descriptions of what would happen to light if they interacted with such materials.

Metamaterials are made from relatively ordinary substances such as fibreglass, copper, silver or other metallic compounds, but the ingredients are built into intricate mosaics of repeating patterns. They can interact with electromagnetic waves in a way that no natural material can, for example creating a surface with a refractive index that is less than 1. This leads to some strange properties – light entering a metamaterial will turn the wrong way, as if it had bounced off some invisible mirror after passing the air–metamaterial boundary. There have been some notable successes: in 2006, scientists at Duke University demonstrated an invisibility cloak that deflected microwaves, the same wavelengths used for radar. Normally the microwaves would have bounced off the material but, instead, they split and flowed around a metamaterial cylinder (which was engineered to have a range of refractive indexes from 0 to 1 along its length) and merged back together on the other side. Anything inside the cylinder would be invisible to the radar.

A metamaterial is an intricate structure that can make electromagnetic waves behave in odd ways around it, even making things 'invisible' to some frequencies of light.

This movement of the microwaves is similar to the way river-water flows around a rock. If you were standing downstream and out of sight of the rock, the pattern of water waves where you were would not tell you that there was a rock further up the stream. It would be invisible to you. Making objects invisible to radar is one thing, making them invisible to our eyes is more difficult. Radar waves are a few centimetres long, and the metamaterials that can manipulate them need to have features a few millimetres across. Visible light is thousands of times smaller than radar waves, requiring metamaterials with an even finer structure. Never to be outdone by such things as technical limits, however, in 2008 Xiang Zhang at the University of California, Berkeley, made a small object 'disappear' from view when it was surrounded by a ring of metamaterials made from silver and magnesium fluoride with a nanometre-scale fishnet structure. Scaling these discoveries up will be difficult for sure. And, at first, it will no doubt be very expensive and limited probably to military or intelligence uses. But we all know that technology has a way of confounding anyone who wants to predict it, so who's to say that buying your own personal invisibility cloak some time later this century will be impossible? Watch this space.

8 How to put the world in order

- Elements
- The periodic table
- The beginnings of chemistry
- The small role of the nucleus
- The table today
- The bonds that make the world
- Finding your way around the elements

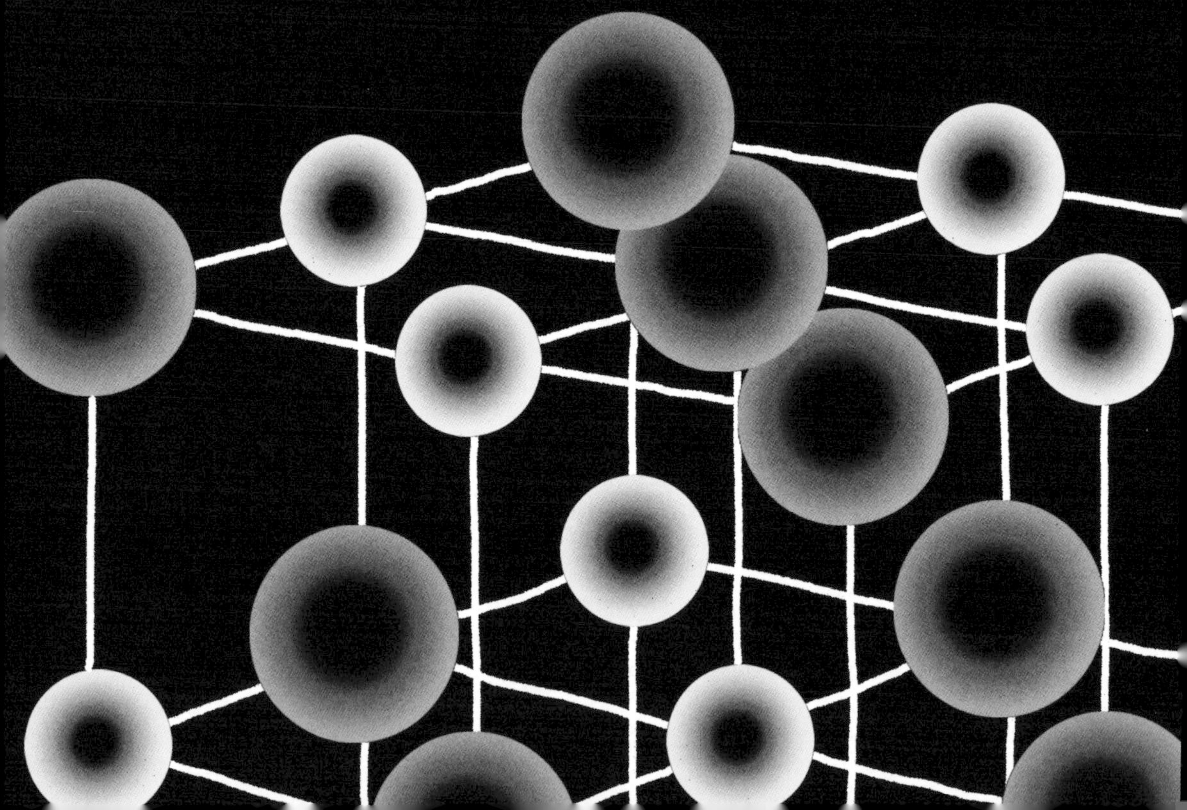

Everything we know of in the world (and everything we have so far been able to detect to the edge of the Universe) is made of one of 94 naturally occurring elements. You need just 25 of them to make a human body and just one – carbon – to make a diamond. Each element has its own chemical and physical properties, they look different, react in a myriad web of interactions and have individual uses and toxicities. From a distance, all this variation is a mess: how can it be possible to get a handle on the diversity? How do chemists manage to keep all these differences in their head?

Elements

Atoms contain a bundle of protons and neutrons at their core, surrounded by a buzz of electrons. The core, called the nucleus, contains all of the positive charge in an atom and most of its mass and, apart from one important bit that decides where an element sits on the periodic table, is largely irrelevant to chemistry. All the chemical reactions that we know of, all of the compounds that are made in the Universe, are a result of the interaction between the clouds of electrons around different atoms. You could say that chemistry is simply the study of how electrons behave.

Not just any electrons, though. In a typical neutral atom, there are equal numbers of protons and electrons. But, while the protons sit still at in the nucleus, the electrons orbit the centre of the atom in a series of shells, each shell slightly further from the nucleus, each one a different shape, containing a different number of electrons at slightly different energies. The electrons that interest chemists sit right at the edge of this layering of shells. These are the so-called 'valence' electrons, which can be shared with or donated to other atoms to form chemical bonds. The number of valence electrons that an atom has defines its chemical properties.

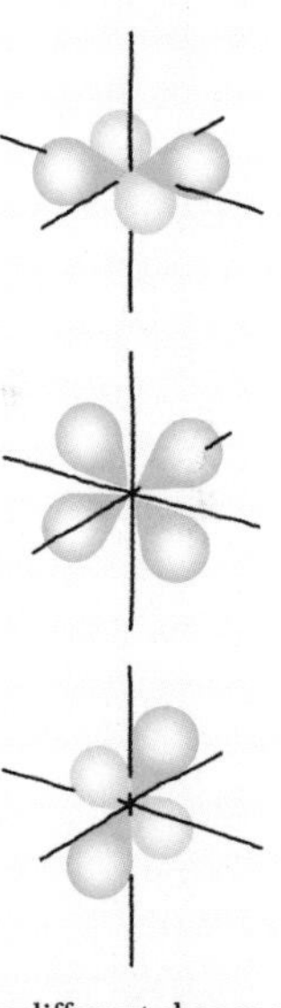

The different shapes of electron orbitals around an atomic nucleus.

The periodic table

This familiar chart, pinned onto classroom walls and printed in chemistry textbooks the world over, is one of the iconic images in science, probably along with the double-helix picture of DNA. Not everyone would know where it came from or even what it means but, as an image, it screams chemistry. The man credited with assembling the periodic table is the Russian chemist Dmitri Mendeleev. In 1869 he put the 65 known elements into order according to their atomic weight (the number of protons and neutrons in their nucleus) and their valency. When he presented the table to the outside world, he showed how it unveiled patterns in the properties of different

1	2	3	4	5	6	7	8	9	10	11	12	13	14	15	16	17	18
H																	He
Li	Be											B	C	N	O	F	Ne
Na	Mg											Al	Si	P	S	Cl	Ar
K	Ca	Sc	Ti	V	Cr	Mn	Fe	Co	Ni	Cu	Zn	Ga	Ge	As	Se	Br	Kr
Rb	Sr	Y	Zr	Nb	Mo	Tc	Ru	Rh	Pd	Ag	Cd	In	Sn	Sb	Te	I	Xe
Cs	Ba	▼	Hf	Ta	W	Re	Os	Ir	Pt	Au	Hg	Tl	Pb	Bi	Po	At	Rn
Fr	Ra	▼	Rf	Db	Sg	Bh	Hs	Mt									
		La	Ce	Pr	Nd	Pm	Sm	Eu	Gd	Tb	Dy	Ho	Er	Tm	Yb	Lu	
		Ac	Th	Pa	U	Np	Pu	Am	Cm	Bk	Cf	Es	Fm	Md	No	Lr	

This is the periodic table of elements. Columns are called groups and contain elements with similar chemical properties. The rows are periods.

elements. When arranged in his table, the elements showed regular 'periodicity' in their chemical properties. Those that had similar atomic weights also exhibited similar chemistry. Mendeleev also remarked that the table could predict the atomic weights and chemical properties of some as-yet-undiscovered elements, since there were gaps in his table below boron, aluminium and silicon.

His original table was not perfect – it did not contain space for isotopes of the elements (atoms of an element that are heavier than the most common version because they have extra neutrons in their nuclei) and he had put some of the elements in the wrong place. But it provided a way to order the atoms and gain an insight into them that had been lacking until then.

The beginnings of chemistry

By the start of the 19th century, chemists agreed that elements were made of atoms that were solid, indivisible and differently weighted, depending on the element in question. Around this time the Manchester-based chemist John Dalton began drawing symbols for the known elements, each one a circle distinguished by a different arrangement of dots, letters, lines and shading. He also tried to work out the relative weights of these atoms by how they combined together into what he called 'compound particles'.

It was already known, for example, that hydrogen reacted with eight times its weight in oxygen to make water. Dalton assumed that an equal number of hydrogen and oxygen atoms must be involved, which gave oxygen a relative atomic weight eight times that of hydrogen. He turned out to be wrong – in fact two hydrogen atoms react with one of oxygen to produce water, giving an oxygen atom a relative weight 16 times that of every hydrogen atom. Though he made mistakes, Dalton did push chemistry into becoming a more systematic and numerical science.

Discovery of other patterns among the elements continued. In 1817, German chemist Johann Döbereiner found that some elements with similar properties fit neatly into groups of three. One of these 'triads' was lithium, sodium and potassium; another was chlorine, bromine and iodine. In the next 30 years, further groupings were found, some with more than three elements – nitrogen, phosphorus, arsenic, antimony and bismuth, for example, which is today labelled as group 15 of the periodic table. Mendeleev built on the work of his contemporaries to construct his table but, though his work hinted at the deeper truth about the nature of what makes an element an element, it would take the discovery of the atomic nucleus to answer that question properly.

The small role of the nucleus

British chemist Ernest Rutherford discovered the atomic nucleus in 1907, when he fired alpha particles at a thin leaf of gold foil. Far from being hard balls, atoms contained most of their mass in a tiny, positively charged nucleus at the centre. A few years later, Danish physicist Niels Bohr completed the 'solar system' model of the atom, showing how he thought electrons must orbit the nucleus. We now know that Bohr's model was too simplistic, but it still works as an analogy for what is going on in an atom.

Mendeleev knew nothing about the structure of atoms when he composed his table but Rutherford's discovery quickly showed that it was the atomic number (the number of protons in a nucleus) that decided what an element was, rather than its atomic weight (the total number of protons and neutrons). His periodic table made much more sense when it was ordered by atomic number.

The table today

In the century after Mendeleev first proposed it, the periodic table went through hundreds of versions. There were arguments about the correct shape for the matrix (some suggested circular or even three-dimensional tables) and new elements had to be fitted in as they were discovered. It might seem part of convention today but the common form of the periodic table, the one we know today, was only formally agreed by the International Union of Pure and Applied Chemistry in 1985. Chemists today know that there are at least 118 elements – 94 natural ones and 24 made synthetically in laboratories and particle accelerators. The elements in the table increase in atomic number along rows, called periods, from left to right. At the end of a period, you move back to the left hand side of the table, this time one row down.

'We are star stuff which has taken its destiny into its own hands. The loom of time and space works the most astonishing transformations of matter.'

Carl Sagan

The atomic weight of an element is the total number of protons and neutrons in a nucleus. In the lighter elements, there are roughly equivalent numbers but the number of neutrons increases with the heaviest elements. This means two elements can have the same atomic weight depending on which isotopes they have. Atomic number – the number of protons – is therefore the better way of telling atoms of different elements apart. It also tells us how many electrons an atom has, and it is the electrons that matter to chemists.

The bonds that make the world

Molecules are formed from collections of atoms bound together by their electrons. Compounds almost always have completely different properties from their constituent elements – water is a liquid at room temperature while hydrogen and oxygen are gases. The development of quantum mechanics in the 1920s destroyed any ideas that chemists once had about Bohr's neat solar-system arrangement of electrons around a nucleus. Nowadays, they talk of 'shells' of electrons where electrons of different energies exist. When an atom gives up an electron, it becomes a positively charged ion, which is then attracted to negatively charged ions. When they meet, they form an 'ionic bond' in a ratio neutralizing the overall charge of the resulting molecule. Sodium chloride is made from a sodium ion with a single positive charge and a chloride ion (a chlorine atom with an extra electron) with a negative charge.

Carbon and silicon, in group 14, form a different type of bond. They have four outermost electrons, so they tend to prefer bonding with atoms by sharing, rather than donating, electrons. In such a 'covalent bond', a single electron is effectively filling the shells of two atoms at once. Covalent bonds between atoms can involve up to three pairs of electrons. Some elements, such as carbon, have sophisticated covalent bonds with a ring of atoms that can share their valence electrons. The type of bonding determines how a compound behaves or what it can be used for. An unhealthy saturated fat differs from a healthier unsaturated fat because the carbon-based molecules in the latter have at least one bond that uses two pairs of electrons or more. In a saturated fat, the carbon bonds only use single pairs of electrons, meaning that the molecules are fully 'saturated' with hydrogen atoms.

Finding your way around the elements

The columns in the periodic table are called groups. In group 1, the alkali metals (lithium, sodium and potassium) all have one electron in their outermost shell that, if it was given up, would leave atoms with a full shell of outer electrons, albeit leaving the atom with a positive charge. In group 17, the halogens (fluorine, chlorine and bromine) all need another electron to fill up their outermost shells. Both these groups are therefore highly reactive with each other (sodium chloride, or common salt, is an example) and with many other elements.

Groups 3 to 12 are the transition metals. In many ways they are all very similar, producing coloured salts and able to lose various numbers of electrons from their outer shells. As their atomic number increases they tend to get less reactive and, at some point, radioactive. This means that the nucleus is too big to be stable and will eventually decay into two smaller atoms, releasing some energy in the process in the form of radiation. At the end of the table, group 18, are the noble gases. They already have full outer shells of electrons and therefore tend to be uninterested in any chemical reactions, making them inert.

Mendeleev's table is still an actively changing thing. There are problems with it: where does hydrogen go, for example? It has one electron (by definition in its outermost shell) so it should fit above lithium in group 1. But it shares none of the properties of the alkali metals. It is more similar to fluorine but, again, hydrogen gas misses many of the properties of halogens. As such, chemists have designed a fudge – giving hydrogen its own place floating in the middle of the table.

The table also keeps growing as new elements are discovered. In 2009, element 112 was given the name 'copernicium' in honour of the Renaissance Polish astronomer, Nicolaus Copernicus, and gained its rightful place in the expanded table. Scientists had made a few atoms of it in a particle accelerator a few decades before but chemists, being cautious, like to take their time to check things out. Similarly, new-elements-in-waiting numbers 113 and 117 could be added soon, after being found in Russian labs. The definitive table of elements is not quite finished yet.

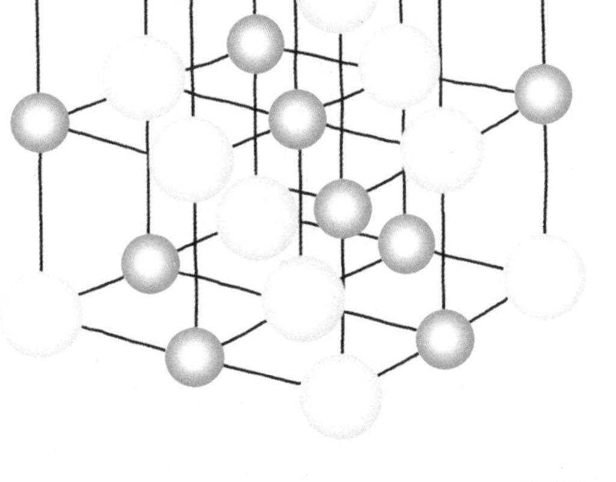

The structure of a sodium chloride crystal (common salt). Dark blue circles are sodium ions; light blue are chloride ions.

9 How to make artificial life

- Right back to molecules
- Biotechnology 2.0
- Welcome to engineering
- Biobricks – biological spare parts
- Technology gets easier and cheaper
- The first artificial life form

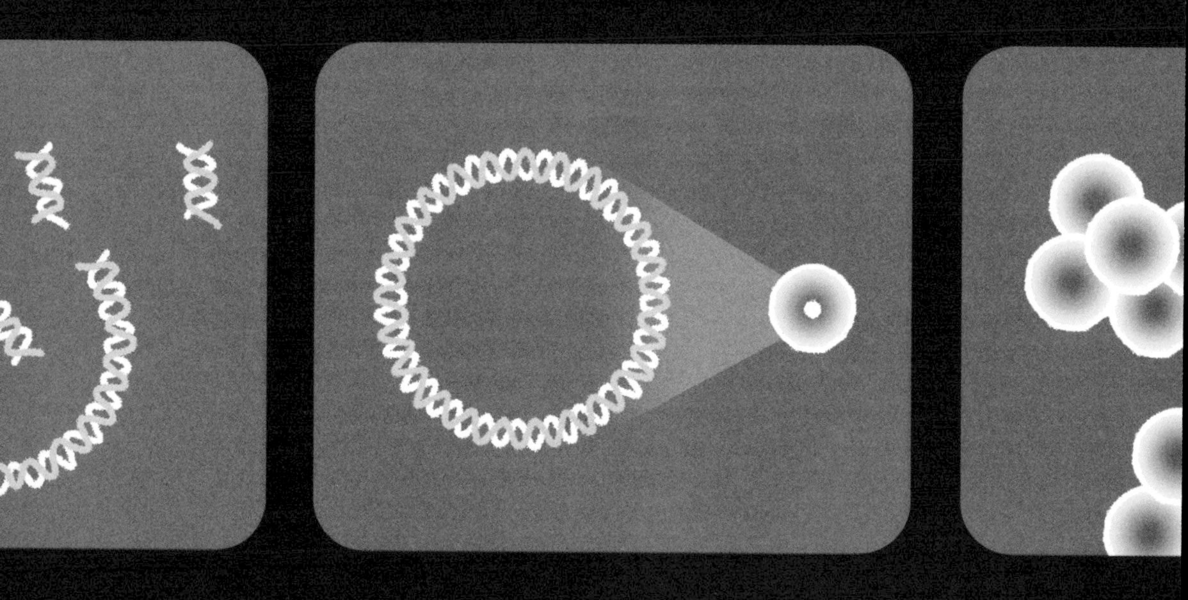

Understanding how changes in DNA bring about new features and new life has been a standard element of biology for decades. But at the end of the 20th century, engineers also got interested. What if you could use the machinery of DNA to mimic some of the basic machinery of life and use it to produce environmentally friendly fuels? This is the promise of synthetic biology, the big new discipline for the 21st century. It promises to open up understanding of how life functions and, along the way, harness the stuff of life itself.

Right back to molecules

Synthetic biology involves developing the tools to build DNA and create reliable methods to manipulate it to do the things you want. The first laboratory-based applications are already here. At the Massachusetts Institute of Technology (MIT), scientists have used viruses to make tiny wires for microelectronic circuits. The viruses are engineered to express proteins and interact with organic and metal compounds to then turn them into long thin wires or rings. No toxic chemicals, high temperatures or environmental issues.

One of the first things that US biologist Drew Endy, one of the leading lights in synthetic biology, created was a counter to process information. Far from being intended to replace the circuits in your laptop, rather it will implement modest amounts of memory and logic in places where there currently isn't any. Perhaps a biological counter in a liver cell counts the number of times a cell divides. Another biological device could monitor the counter so that if the cell has divided more than 200 times – losing control of cell division and risking becoming a tumour – it is killed. It could be a way to beat cancer.

Biotechnology 2.0

Synthetic biology emerged from the disappointingly slow progress, in engineering terms, of biotechnology. In the half-century since the discovery of the structure of DNA in 1953, biologists have developed robust ways to manipulate genes and find new drugs, but their techniques are imprecise and inefficient. That's why so few drugs succeed and why they end up so expensive – every pill you pop has a cost based on all the failures during development. Genetic engineering is almost random in most cases, taking lots of trial and error to learn what is, in effect, a genetic manipulation. Even then, definitive answers are not easy. Synthetic biology replaces the biological mindset with an engineering one. It uses precise mathematical models to predict the behaviour of DNA or proteins, or some combination of the two. And its applications can extend beyond medicine and health.

Welcome to engineering

Biotechnologists might identify interesting proteins or compounds, but they let chemical engineers figure out how to make them on a large scale. Genetic engineering itself is a misnomer – there is no engineering of the biology going on. Proper engineering involves starting with a precise mathematical model of the system you want to build, whether that is an aqueduct or a microelectric circuit. That model will be tested to work out how it behaves in a range of conditions before the system is built to exact specifications from raw materials, so that it will behave exactly as predicted. Any changes in the performance of an engineered system, a car engine say, can be predicted if the engineer knows the inputs, such as the air-fuel ratio or ignition temperature. This capability has not been available in biology until now, because of the lack of quantitative data needed to build robust mathematical models.

That much-needed raw data began flooding in when genomes began to be sequenced. Biologists teamed up with engineers to start building a systematic understanding of what to do with all the sequences. Where biologists struggled with volumes of information, engineers got excited. They saw that, like chemistry and electronics decades before, sequencing data could be used to build models of potential biological systems.

The information helped biologists to unpick how biological systems worked, and it helped the engineers to interact with, modify and apply the biology. Ultimately, their aim is to manipulate biology just like engineers already manipulate complex physics and chemistry – to bring together components of biology that have never before been seen in nature.

'Well done, Dr Frankenstein. But your great grandfather would've been disappointed with you.'

Biobricks – biological spare parts

If biology is to truly turn into an engineering technology, it needs standard components that can be plugged together with minimum fuss. Some synthetic biologists might need to understand the atomic-level interactions between an amino acid and a part of DNA, but most just need to be able to take it for granted that something works. When he was at MIT, Drew Endy started building the biological parts database, a system of standardization that hides complexity from those that do not need to see it.

Where electrical engineers have transistors, capacitors and resistors, biological engineers have 'biobricks'. The first biobricks were made by a MIT computer scientist with an interest in biology, Tom Knight, and are bits of information specifying a piece of DNA and the function encoded by it. Using a range of biobricks, synthetic biologists will be able to build their own systems. Biobricks come in three flavours: 'parts' encode basic biological functions; 'devices' are made from a number of parts and encode human-defined functions (similar to logic gates in electronic circuits); and 'systems' perform tasks such as counting. Knight made the first six biobricks but, thanks to work by Endy and students around the world, the database has now grown to more than 3,000 parts. It grows every year through competitions to design the most interesting components and build applications using those already in the repository.

At the annual International Genetically Engineered Machine meeting, scientists compete to build machines that could win them a coveted trophy – a metal Lego brick the size of a shoebox. One year, a team from the University of Heidelberg in Germany re-engineered an *E. coli* bacterium into a microscopic suicide bomber, which swam towards harmful bugs and secreted natural toxins, killing itself and the pathogens in the process. The same team has also created *E. coli* that find and destroy cancer cells in mice. Another team from Edinburgh University produced bacteria that could detect arsenic in water, which could be used in a freeze-dried version in parts of the developing world where arsenic poisoning is a problem.

Technology gets easier and cheaper

The ability to read, redesign and build genomes into biobricks would have been a tough task, even at the start of the 21st century. The first human genome took the best part of a decade of work to sequence but nowadays reading DNA is thousands of times faster – the increasing number of genome sequences published in recent years is testament to that. Writing sequences is harder but estimates put that technology only a decade or so behind current sequence-reading technology. For scientists, it is already possible to build DNA sequences easily: specify your sequence of base-pairs online to one of several biotech companies in the business of DNA production, and they can deliver a sequence within a few days. The importance of this step cannot be overestimated. Synthesis technology enables the engineering of biology.

'I want to be able to design and build biological systems to perform particular applications. The scope of material I can work with is not limited to the set of things that we inherit from nature.'

Drew Endy

It means scientists can programme DNA, says Endy. 'Imagine what the science around the origin of the Universe might be like if physicists could construct universes. It just so happens that in biology, the technology of synthesis [allows you to] instantly take your hypothesis and compile it into a physical instance and then test it.'

There are already plenty of ideas for applications. Synthetic biologists want to design viruses that keep the organism's useful properties (such as getting genetic material into a cell) but engineer out the bad parts (inflammation, for example). Given the shortage of human liver donors, scientists have been looking at ways of building new livers in the lab or to knock out the viruses that can attack it. All of this sounds fine in principle, but greater knowledge of how viruses make their way into cells and are then transported through the intracellular traffic is needed before synthetic biology can step in. In a field so loaded with possibilities, it is difficult for anyone to map out the future. But Douglas Lauffenburger of MIT is certain that within 50 years, the entire pharmaceutical industry will operate on an engineered basis, eliminating the need for messy trial-and-error methods of drug discovery.

The first artificial life form

Building biobricks is exciting enough but building a whole life form would be mind-blowing. Putting aside, for now, the ethical issues around creating life in a laboratory, creating life forms programmed for specific functions could be very useful. If anyone looks close to achieving that goal, US biologist Craig Venter is the man. He already has plans to make a variety of bugs that will do useful things, such as produce hydrogen for environmentally friendly vehicles. His dream work, however, is geared towards creating the first artificial life from scratch. In 2008, Venter's team reported in *Science* magazine that it had built the entire genetic code of the *Mycoplasma genitalium* bacterium in the laboratory, a supercharged version of the work synthetic biologists were already doing assembling DNA sequences into biobricks. The next step is to insert this synthetic chromosome into a cell and boot it up to start reproducing. Venter's team has also demonstrated that transplanting the genome of one type of bacterium into the cell of another can change the cell's species, a process his team will use with the synthetic chromosome.

Think of Venter's artificial genome as an operating system for a computer. By itself it does little, but install it on a computer and, hopefully, you have a working computer system. The genome is the operating system for a biological cell and the cytoplasm in the cell (all the cell machinery around the nucleus) is the hardware that's required to run that genome. Designer organisms based on these principles have huge potential for creating alternative sources of energy or tackling climate change by soaking up carbon dioxide. Venter has claimed that an artificial fuel-producing microbe could be the first billion- or trillion-dollar organism: 'When you think of all the things that are made from oil or in the chemical industry, if in the future we could find cells to replace most of those processes, the ideal way would be to do it by direct design.'

But let's focus on those ethical concerns for a moment. Critics of synthetic biology raise the problem that scientists could make dangerous organisms by accident, or that the knowledge could fall into the hands of terrorists who would resurrect long-contained viruses, or design more deadly strains. But it's worth putting synthetic biology into context for now: *M. genitalium*, the only recreated genome so far, has about 582,000 base pairs (genetic letters) in its sequence of 485 genes, compared to 3 billion base pairs for humans in just under 30,000 genes. So there's some way to go before nasty things turn up.

And no technology has ever been developed without possible bad uses. As the bioethicist Tom Shakespeare once wrote: 'Research should progress carefully, and with appropriate safeguards and scrutiny. But we should also be careful about resorting to grandiose claims about either the wonders or the wickedness of life science.'

To create his artificial version of a common bacterium, *Mycoplasma mycoides*, Craig Venter began with a computer reconstruction of the bug's genome.

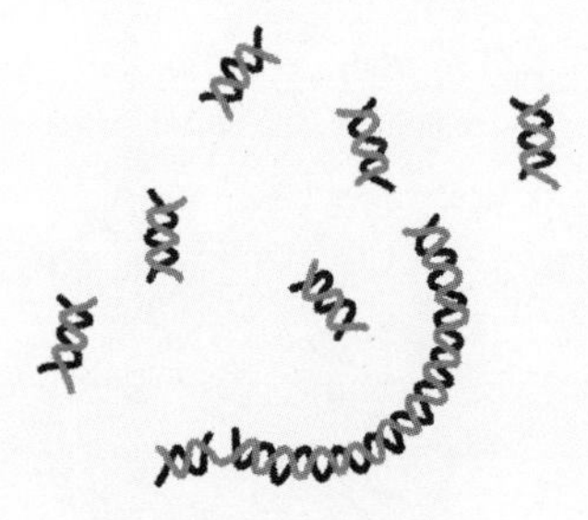

1. A computer reconstruction of the genome is fed into a DNA synthesizer, which produces short strands of the bug's DNA.

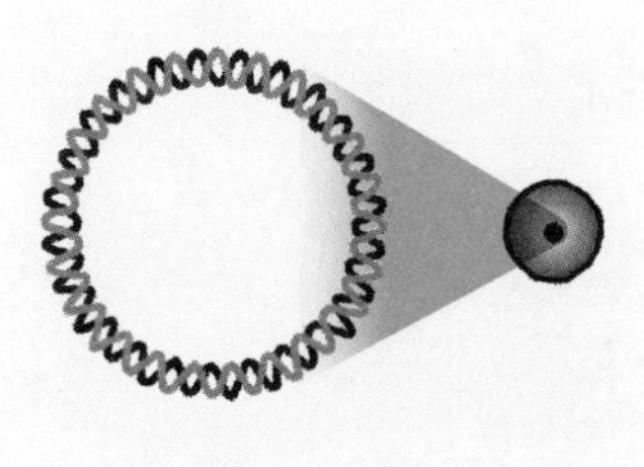

2. The strands are inserted first into yeast and then into E. coli *bacteria, and in the process 'repair' themselves into a circular genome.*

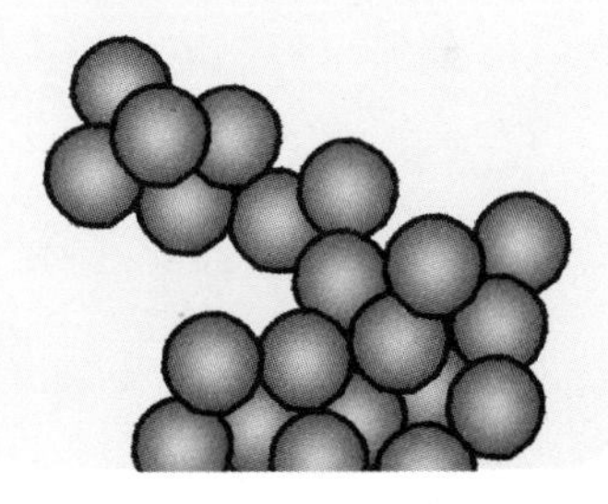

3. This new, synthetic genome is transferred into another bug; its offspring ditch their own DNA and use the synthetic genome.

10 How to build a Universe

- The scale of things
- The first ten seconds
- Then the Universe inflated
- The next five minutes
- From five minutes to 13.7 billion years
- The stars are born

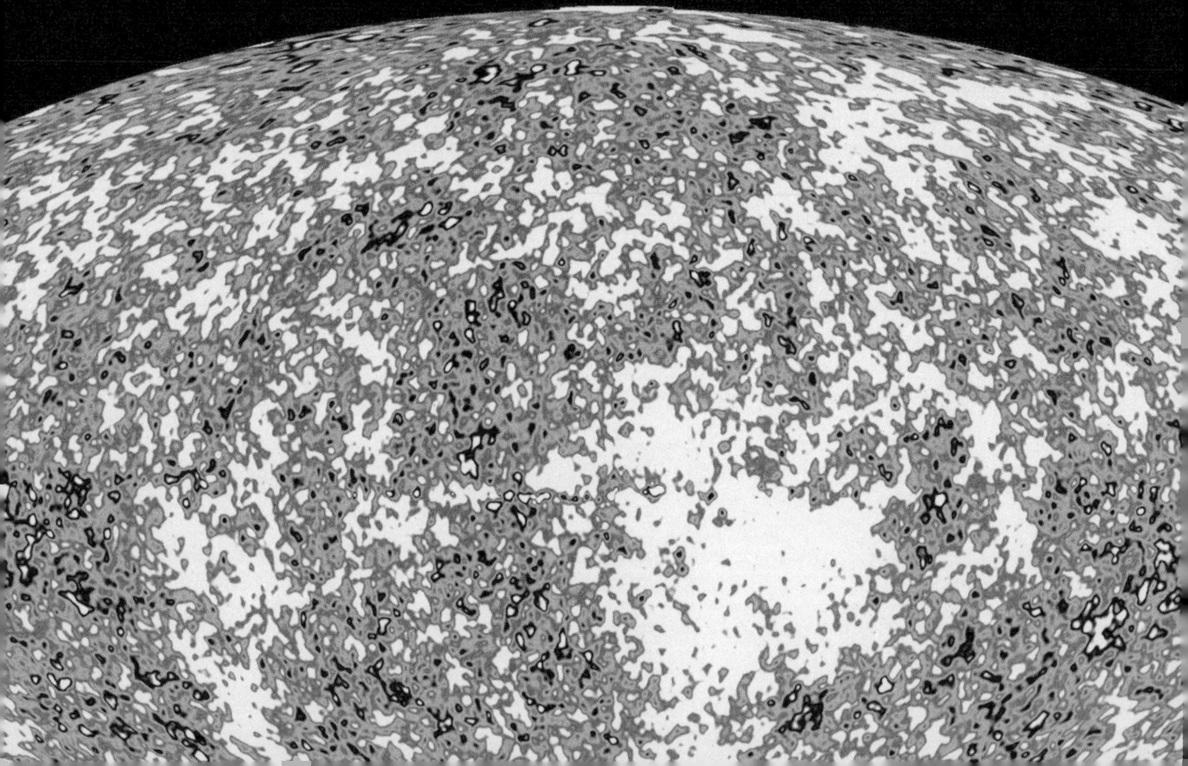

About 13.7 billion years ago, a tiny fraction of a second after the Big Bang, the tiny, infant Universe was pure, furious, incredibly hot energy. Time started, the expansion began, some of the energy became particles of force, then of matter. When things cooled, electrons, protons and neutrons formed simple atoms. The atoms came together to make stars and cooked the other elements while coalescing into galaxies.

The scale of things

The Universe is huge. Getting to grips with the timescales and distances involved isn't easy. Assuming that the light from the farthest objects we can see started its journey in the earliest years of the Universe, 13.7 billion years ago, you might think that the Universe is some 13.7 billion light years in every direction. Not quite. While that light has been travelling to us, the space behind it has been expanding. Cosmologists calculate that the edge of the observable Universe is much bigger than the 13.7 billion light years travelled by the oldest light – it is more like 47 billion light years in each direction. Beyond that, the actual Universe may well be near-infinite in size.

Take a moment to think about those numbers. Our galaxy, the Milky Way, is around 100,000 light years across and our nearest neighbour, Andromeda, is around 2.5 million light years away. In the 47 billion light-year-radius of the Universe there are at least 100 billion galaxies, each containing billions of stars. And all of that stuff, those trillions-upon-trillions of stars and an even more mind-boggling number of atoms and particles, started at a single point, a singularity of infinite density. The timescales are also hard to comprehend. The history of everything started in the few seconds and minutes after that singularity erupted. Then nothing much happened for thousands of years before some of the bigger, more complex objects, such as stars and galaxies, began to emerge.

The evolution of the Universe from the Big Bang (at the extreme left), through a sea of fundamental particles to today (at extreme right) with modern galaxies.

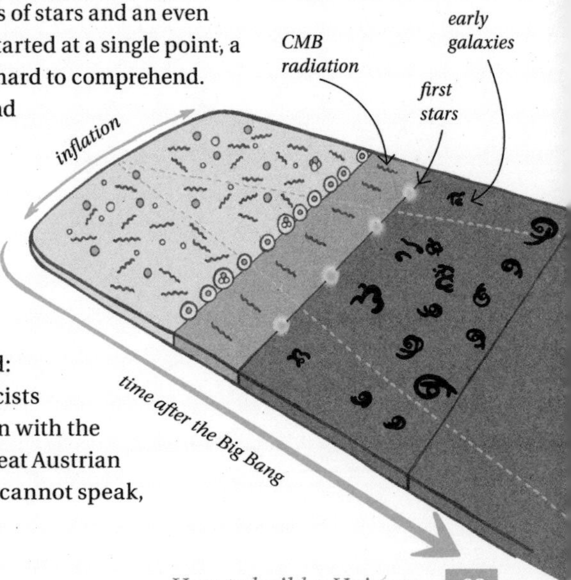

The first ten seconds

Before we start with the singularity at the start of time, we must deal with a question that is often posed: what happened before the Big Bang happened? Physicists answer this question by arguing that time itself began with the Big Bang so there is no such thing as 'before'. As the great Austrian philosopher Ludwig Wittgenstein said: 'Whereof one cannot speak, thereof one must be silent.'

The Hertzsprung-Russell diagram shows the relationship between a star's luminosity (y-axis) and its temperature and colour (x-axis). The diagram shows that stars can exist only in certain combinations. The most dominant is the diagonal line from top left (hot, bright) to the bottom right (cooler, less bright), called the main sequence. At the bottom left of the diagram are dwarf stars. Above the main sequence are the giants. A star will typically move around on this diagram as it evolves through its life and its mass and temperature changes.

Centuries of astronomical observations and decades of experiments in particle colliders have given scientists the ideas they need to paint a picture of the Universe in the moments after the Big Bang. Today they can say what happened up to 10^{-15} seconds after the beginning, but before that we are in the realms of speculation.

It started with a singularity, a point with no dimensions that was of infinite density. Physics as we know it wouldn't apply there so it is impossible to know why it did what it did next: explode.

The period 10^{-43} seconds, known as the Planck era, is thought to be the smallest possible unit of time, and no-one knows whether it makes scientific sense to talk about anything shorter than this period. The one thing of note that happened in the Planck era is that one of the four fundamental forces of the Universe, gravity, separated out of the messy cloud of energy. The Universe was filled with gravitons, the particles thought to carry the gravitational force, and also the particles of the as-yet-unnamed 'unified' force that combined the other three fundamental forces of nature: electromagnetism, the strong nuclear force and the weak nuclear force. The temperature dropped from an estimated 10^{32} K in the instant after the Big Bang to 10^{29} K, still around 22 orders or magnitude hotter than the centre of the Sun. At 10^{-32} seconds after the Big Bang, temperatures had dropped enough for the strong nuclear force (which keeps the bits of a nucleus glued together) to split from the unified force, leaving the electromagnetic and weak forces joined together as the electroweak force.

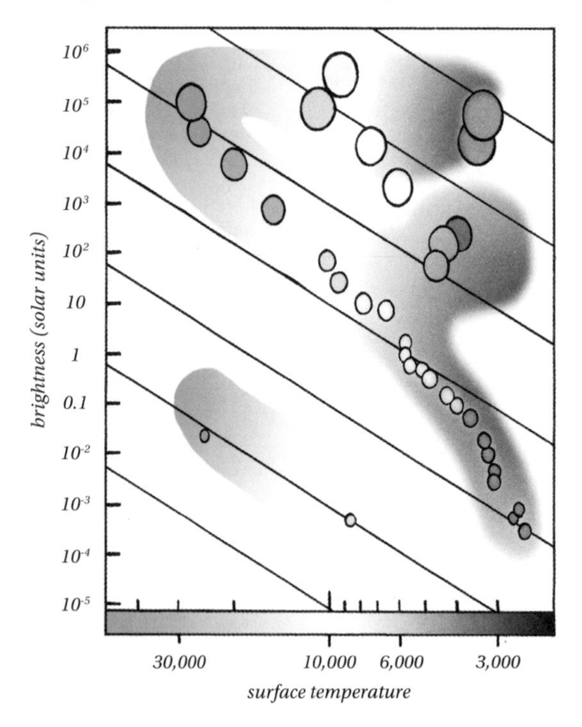

Then the Universe inflated

At some point, the rate of expansion shifted to an astonishing rate, faster than the speed of light. The Universe more than doubled in size every 10^{-35} seconds and, by the time this inflation was switched off at 10^{-32} seconds after the Big Bang, the Universe had done this 100 times. To put that into context, imagine the Universe started off at a scale of 1 cm. After 10^{-35} seconds, one 'tick' of inflation, the Universe would be 2.7 cm wide. Two ticks, it would be 7.4 cm. Three ticks later, we're at 20 cm. By 20 ticks, the Universe is

4,850 km across and, after 50 ticks, 5,480 light years. All that in less than 10^{-33} seconds. By the time inflation had finished, after 100 ticks of inflation, the Universe would have grown by a factor of 10^{43}. And that is a conservative version of the theory – in some theories, inflation was even more extreme, with the factor of expansion more like 10 multiplied by itself a trillion times.

'If you wish to make an apple pie from scratch, you must first invent the Universe.'

Carl Sagan

And the Universe continued to cool. Between 10^{-32} seconds and 10^{-10} seconds, the energy in this suddenly expanded Universe began to crystallize into the particles of matter from which we, and everything we can see in the Universe, are made. This includes the quarks, the constituents of protons and neutrons, and leptons, which include electrons and neutrinos. Six types of quarks and six leptons are the basic matter particles that make up today's Standard Model of particle physics, the acknowledged quantum mechanical description of the world. But the quarks and leptons present in this early Universe were not all that distinguishable. Apart from their masses, the particles were largely interchangeable and the temperature was so high 10^{-32} seconds after the Big Bang that the proto-quarks and proto-leptons all roamed freely. Not that they survived for very long anyway – as well as matter, the early Universe contained lots of antimatter. These are particles with exactly the same mass as their matter equivalent but with opposite electrical charge. Whenever a matter quark or lepton came into contact with its antimatter equivalent, the particles would be annihilated into a shower of high-energy photons. This might have carried on forever, destroying all matter particles were it not for a quirk in the early Universe. For every billion particles in total, there was an extra matter particle. Those are the quarks and leptons that survived and we are all made from this tiny leftover of the violent annihilations.

At 10^{-10} seconds old, the remaining two fundamental forces emerged in the Universe. The electroweak force (the remnant of the 'unified' force after the strong nuclear force had separated out) split into the weak nuclear force, which is responsible for radioactivity in the nucleus of atoms, and the familiar electromagnetic force, the most relevant force for all of our daily experiences on Earth – everything from chemistry to electricity. For the rest of time, the four forces would maintain different identities. A millionth of a second later, the temperature had dropped enough for the quarks to start forming the protons and neutrons that make up the nuclei of all the atoms in the Universe. The neutrinos stopped interacting with quarks and leptons, retreating into the background of the Universe as a ghostly fog of particles that would roam unimpeded throughout the cosmos forever more. The few of the heaviest leptons that had survived being annihilated by their anti-particle equivalents decayed into ordinary electrons.

The next five minutes

At a tender ten seconds old, after so much activity and chaos, the Universe's evolution slowed down. Protons (and an equal number of electrons) outnumbered neutrons by five to one, but all of these particles were drowned in a sea of energetic photons – particles of light – which also transmitted the electromagnetic force around the Universe. There were 1 billion photons for every matter particle in the Universe at this stage. For the next 3 minutes, the photons were so energetic in the hot Universe that they knocked matter particles around, refusing to let them coalesce in any meaningful way. The matter did not have to wait long because, two minutes later, temperatures had dropped to a billion kelvin (K), low enough for two neutrons and two protons to come together to form a helium nucleus. In these few minutes, the nuclei of the first elements emerged: the Universe was 76 percent hydrogen (single protons), 24 percent helium (two protons and two neutrons), with minute amounts of deuterium (a proton and a neutron), helium-3 (two protons and a neutron), lithium-7 (three protons and four neutrons) and beryllium (four protons and a few neutrons). For millions of years, these were the only elements.

From five minutes to 13.7 billion years

For 10,000 years, these nuclei bumped around in the Universe – which looked like an opaque cloud because the photons of light could never travel very far without being scattered into a different direction. It would take another 380,000 years, with temperatures at 10,000 K, for the Universe to became transparent. The oldest light waves that we can detect come from this time – anything before that is trapped in the opaque cloud.

As the Universe expanded, so did the wavelength of the photons. Because the energy of a photon is inversely proportional to its wavelength, the photons soon became too low in energy to keep electrons away from the nuclei that float around and, for the next few thousand years, these particles began to find each other and began making atoms. Over the same time, the photons' energy kept dropping and eventually they stopped interacting with the atoms altogether. After this 'decoupling' event, the low-energy photons take a back seat in the Universe to form the cosmic microwave background, a remnant of the hot ball of plasma that has cooled to 2.7 K above absolute zero. Then, for millions of years, things were calm. Because of the density of the Universe, the nuclear and electromagnetic forces had ruled the Universe, with repulsive electrical forces keeping matter from coming together for too long. The Universe was a cold, dark, featureless place.

The stars are born

As soon as the overall density of the Universe dropped to the right level, gravity took over. Because all the atoms in the Universe felt the effects of gravity, and because gravity is always attractive (so it cannot cancel itself out over large distances like the electromagnetic force) it made atoms clump together. Wherever there was a randomly large amount of hydrogen, gravity condensed and compressed the cloud of gas to such an extent that the atomic nuclei at the centre of the cloud began to fuse. Nuclear fusion released energy and the star began to shine. Our star, the Sun, was made like this.

Stars came together in their billions to form galaxies of a myriad shapes, most of them held together by the gravity of the mysterious dark matter. Stars also became the nurseries for all of the remaining natural elements we know about. When a star has fused all of the hydrogen at its core into helium, the internal pressure of the core will prevent it from collapsing any further. Its outer layers expand and cool to form a red giant some 200 times bigger than the original star. A thin layer outside the core of the star will continue to fuse hydrogen and, eventually, compress the core enough to fuse the helium that is there into carbon and oxygen.

After the helium in the core is all spent, helium fusion continues in a shell around a hot core of carbon and oxygen. Depending on the size of the star, more fusion stages follow, creating heavier elements: successive phases are fuelled by neon, oxygen and silicon. By the time the star is producing iron at its core, it is near the end of its life. Fusing iron is no good for a star since it would consume, rather than release, energy. When the iron core of a massive star gets too big, it collapses. If the star is massive enough, in its dying moments it will rupture space itself and create a black hole, a singularity with such gravitational strength that nothing can escape once it gets close. If it is a small star, the ashes will turn into a ball of ultra-dense neutrons or some other exotic, unrecognizable matter. Either way the star's collapse, after billions of years of fusing elements together, will cause a shockwave that makes it explode into a supernova, an event so bright that it can briefly outshine an entire galaxy of stars. It is from the remnants of these violent supernovae that planets, and ultimately life, are made – but that is another story.

Spanning the entire sky, the cosmic microwave background radiation would be discovered by astronomers in the 20th century and used as one of the many lines of evidence proving that the Big Bang occurred.

11 How to find ET

- Listening to the skies
- The waiting game
- What might alien life look like?
- Universal vs parochial
- What about intelligence?
- Are there clues on Earth?
- Exploring the cosmos
- What are the odds?

The hunt for intelligent species outside Earth is a staple of science fiction, an articulation of our desire to know whether we are alone in the Universe. There are billions of stars in our galaxy and, it is reasonable to expect, an even greater number of planets orbiting them. Some of them, surely, will have a perfect temperature and distance from their star, which would make life possible. Going on sheer numbers alone, it is not unreasonable to expect that some of this life is intelligent and capable of interstellar communication.

Listening to the skies

In 1977, astronomers in Ohio saw something remarkable. Examining the data from one of their radio telescopes, they saw a signal that stood out from the noise around it. Over 72 seconds, it gradually rose and then fell. Jerry Ehman, a professor at Franklin University in Columbus, circled a particular sequence of numbers and letters and wrote 'wow' in the margin. What made this signal interesting was that it came from the Big Ear observatory which, for several years, had been scanning the skies for signals from alien intelligence. In all that time, it had found nothing out of the ordinary. Had the 'wow signal' suddenly changed that? Had Ehman found evidence for extra-terrestrial life? Several weeks later, astronomers pointed the telescope towards the source of the wow signal, hoping to catch it again, to confirm that this really was something to get excited about. But they found nothing. They found nothing again several weeks later, and then several years later, when other astronomers repeated the scan with different telescopes. Eventually, the wow signal was dismissed. And the search for intelligent extra-terrestrial life went on.

The Big Ear observatory was part of a larger project that had started in 1960, when US astronomer Frank Drake pointed the Green Bank telescope towards the star Tau Ceti. He was trying to pick up radio signals that might be hurtling across space, sent by intelligent aliens far from our Solar System. It was the beginning of what would eventually become known as the Search for Extra-Terrestrial Intelligence (Seti), a grand project to find incontrovertible evidence that we are not alone in the cosmos. But, so far, Seti has not been successful. After 50 years of scanning the skies, nothing has even come close to 1977's wow signal. For half a century, the skies have been silent.

The waiting game

When he grappled with the question of ET, Italian-US physicist Enrico Fermi asked why, if life was common in the Universe, we had not been contacted already. Where are they, he wondered. Fermi's question still looms large.

If aliens do exist, if they have had billions more years than humans to evolve and develop technology, why are they not here?

There are lots of practical problems involved in hunting for aliens. Chief among them is distance. Our galaxy is mind-defyingly big. It would take a beam of light 100,000 years to cross from one end to the other. If our nearest neighbours were life forms on the forest moon of Endor, 1,000 light years away, it would take a millennium for us to receive any message that they might send. A response would take the same amount of time to reach the aliens. It is not a timescale that allows for quick banter.

The top image is a possible pattern from aliens. The dots and dashes are transmissions of different durations. Double length dashes indicate the end of a line. The middle image is arranged so that the longer dashes form the end of a row. The bottom image is a line drawing made by connected shorter dashes and a smooth line.

And they might not be communicating in our direction anyway. If the Endorians were watching us, the light reaching them at this very moment from Earth would show them our planet as it was 1,000 years ago. In Europe that means lots of fighting between knights around castles and, in North America, small bands of peoples living on the great plains. If our nearest aliens are tens of thousands of light years away, they would see only the ancestors of modern humans living among much greater beasts. You would forgive them for not bothering to get in touch. A more dire thought is that perhaps aliens have existed but, ever lustful for resources and power, they have destroyed themselves after developing powerful weapons, such as the nuclear bomb. Or is the rest of the galaxy keeping its distance, worried about humanity's violent tendencies?

What might alien life look like?

The lack of a signal from ET has not stopped astrobiologists from coming up with ideas for what properties aliens might have. Throw out any assumptions that they have to resemble the humanoid aliens from *Star Wars*, even with their assorted tentacles, tails or extra heads. The hunt for planets outside our Solar System has focused on 'Earth-like' planets in the hope that one of these perhaps has identical conditions to our own planet and, therefore, a greater chance of life just like ours. But why should life be anything like us? Humans evolved on a planet rich in oxygen and water, where a carbon-based molecule called DNA became the copying mechanism for life. From our point of view, we seem to exist in a world with just the right temperature, water and nutrients.

Aliens, of course, are not restricted to our point of view. You don't need to step off Earth to find life that is radically different from our common experience of it. Extremophiles are species that can survive in places that

'Lots of people think that because they would be so wise and knowledgeable, aliens would be peaceful. I don't think you can assume that. I don't think you can put human views onto them and that's a dangerous way of thinking. Aliens are alien. If they exist at all, we cannot assume they're like us.'

IAN STEWART

would quickly kill humans and other 'normal' life forms. These single-celled creatures have been found in boiling hot vents of water thrusting through the ocean floor or at temperatures that are well below the freezing point of water. The front ends of some creatures that live near deep sea vents are 200°C (360°F) warmer than their back ends. In our naïve and parochial way, humans have named these creatures extreme but, from the point of view of a creature that lives in boiling water, we are the ones who are extreme.

On Earth, life exists in water and on land but, on a giant gas planet, it might exist high in the atmosphere, trapping nutrients from the air. Faced with such diversity, it is still possible to make an educated guess at what aliens might be like. For a start, says Ian Stewart of the Mathematical Institute at the University of Warwick, you have to carve up biological features into things that might be universal across all life forms in the galaxy, and those parochial to Earth.

Universal vs parochial

Parochial features include anything unique to a single species, such as the five fingers on a human hand. There is no reason, other than a quirk of evolution, why we don't have four or even six digits. The eye, for example, has evolved more than 40 times in completely unrelated creatures. But having an organ to see with or limbs to move around with are probably all universal features. As Ian Stewart put it: 'Limbs have evolved independently in different creatures – an octopus has tentacles but they do the same job, but with very different structure.' DNA is most likely parochial to Earth, but the idea of evolution of species by natural selection is probably universal.

What about intelligence?

Intelligent life on Earth is not restricted to humans. There are many intelligent creatures – octopuses, dolphins and whales are bright. Even mantis shrimps are surprisingly good at solving puzzles if they have to get to their food. But intelligent life alone would not be enough for us to detect aliens on another planet. They would need an ability for mass cooperation, and to be able to develop technology. 'What would give us the possibility of communicating with other planets is not intelligence as such, it's our ability to store anybody's bright ideas in a form that the rest of the culture can access and use,' says Stewart. Humans today are individually no smarter than previous generations for a good many thousands, or possibly hundreds of thousands of years. But, collectively, our culture can achieve things that were inconceivable 100 years ago. 'Extelligence', as Stewart calls it, started with the invention of speech and writing, got going with printing, and is now running riot with the Internet. Once a species is extelligent, lots of things become possible – they might transcend biology. Humans went from using radio waves for communication to launching spacecraft in less than 70 years; in 100 years we might have cracked artificial intelligence. Intelligent biology can engineer its own successor, and very quickly, in terms of interstellar communication and travel. We have already gone beyond the confines of a three-pound brain sitting in a skull on top of your neck.

Are there clues on Earth?

Paul Davies, a British astrophysicist at Arizona State University, wonders if we are starting our search for ET in all the wrong places. Perhaps there is 'alien' life already here on Earth. There is no agreed theory for the start of life on Earth so trying to calculate the number of life forms on other planets is a pointless way of working out if life really exists. 'There might be a way to solve this problem at a stroke,' he writes. 'No planet is more Earth-like than Earth itself, so if life really does pop up readily in Earth-like conditions, then surely it should have arisen many times right here on our home planet? And how do we know it didn't? The truth is, nobody has looked.'

'Ok then ... please remember when we land on our Human Nature Reserve, keep yourselves out of sight.'

All life forms use the same DNA to replicate, so it has always been assumed that all life on Earth is somehow related, spawned from a single cell that existed 3.8 billion years ago. But what if there was a creature somewhere on Earth that had a different biochemical signature? 'It is entirely possible that examples of life as we don't know it have so far been overlooked,' writes Davies.

Exploring the cosmos

The hunt for ET and related pursuits, such as looking for planets outside our Solar System, stepped up a gear in the late 1990s and early 21st century. Scientists discovered

the first extrasolar planets in the early '90s but, thanks to better telescopes and improved detection techniques, confirmed numbers of planets have shot up. Today, scientists recognize 443 planets orbiting around more than 350 stars. Most of these planets are gas giants in the mould of Jupiter, the smallest being Gliese 581, which has a mass of 1.9 Earths.

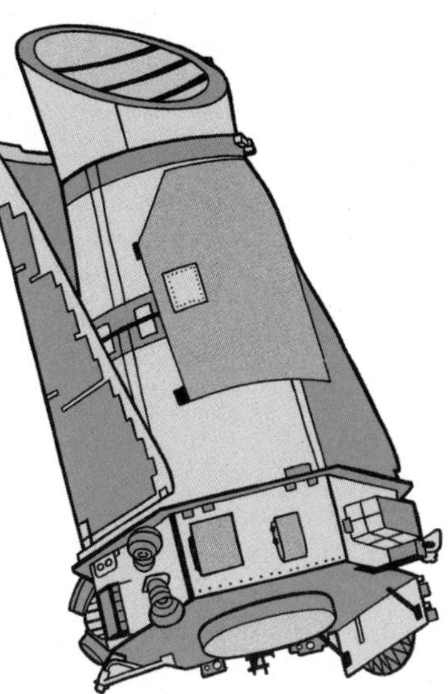

NASA's Kepler probe, launched in 2009, is a space observatory designed to look for Earth-like planets outside the Solar System.

In 2009, NASA launched the Kepler satellite, a probe specifically designed to look for Earth-like planets. By looking at a small patch of the sky for more than three years, the satellite watches thousands of stars for periodic dimming, the tell-tale sign that a planet might have passed in front of it. Future generations of ground-based telescopes, such as the proposed European Extremely Large Telescope (with a 30m (100ft) main mirror) could be operational by 2030, and would be powerful enough to image the atmospheres of faraway planets, looking for chemical signatures that could indicate life.

And, for so long lacking the significant funding it needed to fully sweep the skies for radio waves, the Seti Institute finally has a serious piece of kit under construction: the Allen Array. At present it has 42 radio antennae, each 6 m (20 ft) in diameter, but there are plans to have up to 300 radio dishes. In all the years that Seti has been running, it has only managed to carefully look at less than 1,000 star systems. With the full Allen Array, they could look at 1,000 star systems in a couple of years.

What are the odds?

At a Royal Society seminar on astrobiology in 2010, Frank Drake was the toast of the event, admired as much for his calm and manners as for his dogged patience in getting his ideas taken seriously by the scientific establishment. On the question of whether humans would ever hear that signal that tells us we are not alone, he had lost none of his enthusiasm, even after half a century of silence. 'Fifty years ago I was naive in thinking we could find signals straight away. For all I knew there were radio broadcasts pouring from civilizations on every star,' he told the assembled scientists. 'But that was really unreasonable. I now realize that it is going to be harder than that. There may be up to 10,000 civilizations in the galaxy but, given that the galaxy also contains 100 billion stars, that means we will have to search around 10 million stars before we have a realistic chance of finding one. That is certainly not going to happen in my lifetime. Nor might it happen in the next generation. But we will make contact one day. I am sure of that.'

12 How to join up the Universe

- Electromagnetic force
- Strong and weak interactions
- Gravitation
- Unification of the forces

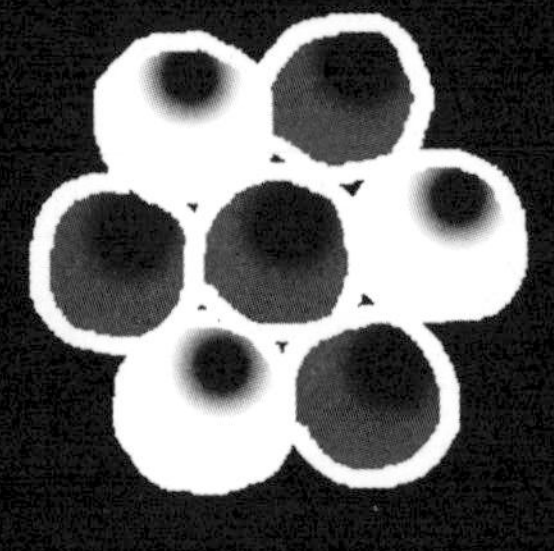

Gravity pulls you down. The north poles of a magnet push each other apart. Electricity speeds down wires to power everything in your home. The Universe is made of stuff: protons, electrons, neutrons and dark matter and dark energy. How all of that stuff moves around, stores and releases energy, and how some of it forms the atoms from which everything we know about is composed, is determined by four fundamental forces, born just after the Big Bang: electromagnetism, a strong and weak nuclear force and gravity.

Electromagnetic force

This is the interaction that shapes everything it is to be human. All of our molecules were shaped by the electromagnetic attraction and repulsion between different atoms and molecules that created and shaped our proteins. The food we eat is metabolized using electromagnetism-based chemical reactions to release energy stored in the electromagnetism-based atomic bonds in them. The reason this book does not fall through your hand when you read it or the table it is resting on is thanks to the electromagnetic repulsion between the electrons in the pages and those in your fingers. The light reflecting off the page and into your eyes is an oscillation in the electromagnetic field that pervades all of space – your eyes interpret different frequencies of oscillation as colours. Your brain and nervous system use electrical currents to process information and communicate with the rest of your body. And we haven't even started on electronics, computers or lightning.

At its core, the electromagnetic force is an interaction between electrically charged particles, such as protons, electrons or ions. It is carried between these matter particles, or fermions, by the particle of light, the photon. It has an infinite range and is far, far stronger than gravity. Take two bricks placed next to each other on a table. The total repulsive force between the trillions of negatively charged electrons in one brick and the trillions of negatively charged electrons in the other brick is bigger than the entire weight of Earth. By rights, these bricks should fly apart at unimaginable velocity. Never mind flying off the table, which they should also be repulsed by. But neither brick nor table move because the electromagnetic force attracts as well as repels. The protons in one brick attract the electrons in the other and vice versa, exactly cancelling out the repulsive force between electrons.

The basic concepts behind electricity and magnetism have been known since ancient times, but it took the physicist James Clerk Maxwell, building on the work of Michael Faraday in the late 19th century, to bring the two

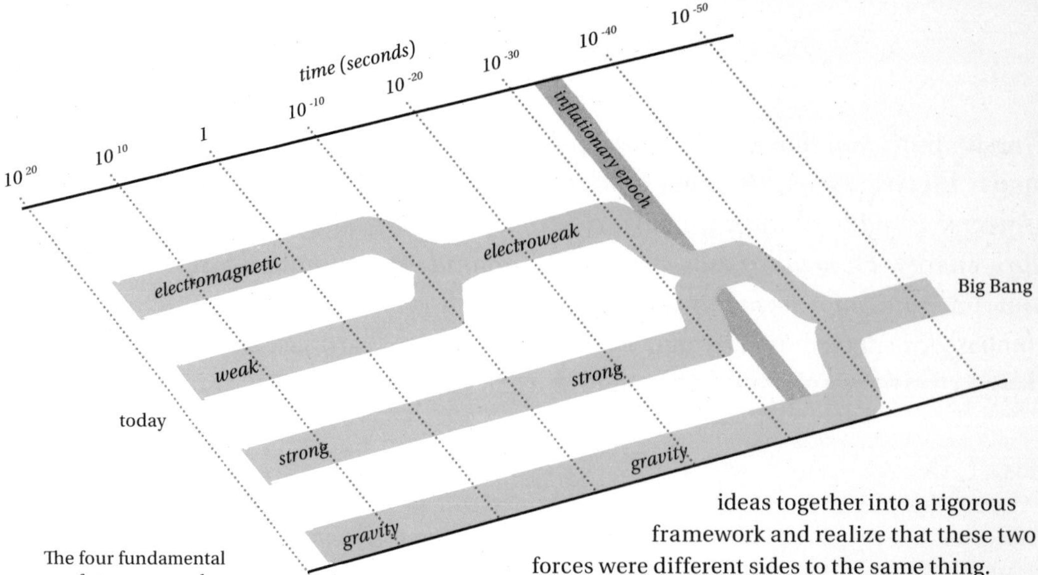

The four fundamental forces emerged at different times after the Big Bang, as the Universe cooled and the particles within it had less energy.

ideas together into a rigorous framework and realize that these two forces were different sides to the same thing. Wherever you can measure an electrical force, as it oscillates down a wire for example, there is also magnetic force, acting at right angles. Maxwell's ideas allowed the electromagnetic force to be harnessed and controlled to make electrical currents.

The force was brought into the strange world of quantum mechanics in the middle of the 20th century, when physicists developed it into a theory called quantum electrodynamics (QED). This precise mathematical description allowed scientists to manipulate the force in unexpected ways, ushering in semiconductors and electronic devices such as transistors, which would be impossible to build under Maxwell's 'classical' electromagnetic theory.

Strong and weak interactions

Deep in the heart of every atom is the nucleus, made of protons and neutrons. And here also lie two more of the fundamental forces which, were it not for cosmologists and high-energy particle physicists, we would have no clue ever existed. Indeed, even physicists did not have a clue these forces existed for a long time. While they had already come up with good models for gravity and electromagnetism, they had no inkling that the nucleus held two more forces, each one fundamental. Partly it was because no-one knew nuclei existed until Ernest Rutherford discovered them after firing alpha particles at a thin sheet of gold. Once that happened in 1908, scientists began to wonder how the particles stayed bound together. The protons and neutrons would have some gravitational attraction between them, for sure, but that would pale into nothing compared with their mutual electromagnetic repulsion. What was it that overcame this mutual repulsion and which operated to keep the nucleus so tightly bound up?

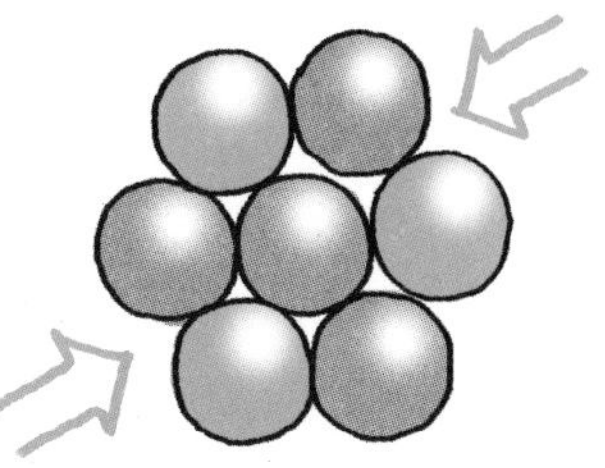

Strong force binds the nucleus

It took a further 60 years after Rutherford's discovery before a suitable theory arose to explain the problem. US physicists Murray Gell-Mann and George Zweig proposed that protons and neutrons were made of even more fundamental particles, called quarks. These held two types of charge: an electrical charge and a colour charge. Despite being repelled from each other by their electrical charges, quarks were attracted to each other by the strong nuclear force, mediated through the colour charge. The theory of quantum chromodynamics (QCD) emerged to explain the strong nuclear interaction in terms of quantum theory, analogous to QED for electromagnetism. Quarks interact with each other via a force-carrying particle called a gluon. How strong the attraction is depends on the size of the colour charge held by each quark.

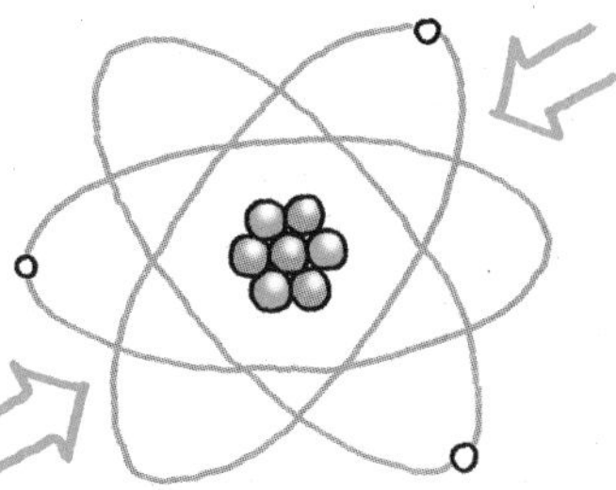

Electromagnetic force binds atoms

The strong force has some odd properties that explain why it is never seen outside an atomic nucleus. The further apart two quarks get, for example, the stronger the gluon force between them becomes, a bit like how a rubber band gets harder to stretch as it gets longer. For this reason, quarks and gluons never exist by themselves and have never been spotted in isolation.

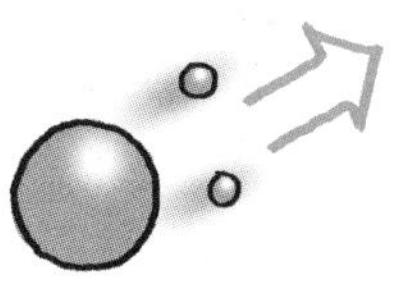

Weak force is responsible for radioactive decay

The weak nuclear force also acts on the quarks. Transmitted by particles called the W and Z bosons, it is responsible for radioactive beta decay in the nucleus, when a neutron decays into a proton and releases an electron in the process. In the late 1970s, scientists confirmed that, at very high energies, the weak force becomes indistinguishable from electromagnetism. Pakistani physicist Abdus Salam and US physicists Sheldon Glashow and Steven Weinberg were awarded the Nobel Prize in Physics in 1979 for their work on the 'electroweak theory', the first unification of forces since Maxwell showed that electricity and magnetism were the same force more than a century earlier.

Gravitation

Gravity is the weakest of the forces. Anything with mass will attract anything else with mass via gravity – in this instance, mass could be thought of as a particle's gravitational 'charge' – it is infinite in range, and it always attracts. The mutual attraction between you and Earth keeps you glued to the ground, draws rocks back to the ground when they are thrown up, and is the force we fight when flying. Gravity is 10^{36} times weaker than the electromagnetic

Gravitational force binds the Solar System

'You will have to brace yourselves for this – not because it is difficult to understand, but because it is absolutely ridiculous: All we do is draw little arrows on a piece of paper…'

Richard Feynman

force but, over cosmic distances, its key property comes into its own. The electromagnetic force (which is also infinite) will cancel itself out. But gravity is never repulsive – it brings things closer together.

For large bodies in space, the electromagnetic force is largely irrelevant in how they move because they contain equal positive and negative electrical charges, in the form of protons and electrons. Whereas gravity will cause all of the particles to attract all of the others, which can become significant when there is a lot of mass involved. As such, gravity is the only truly cosmic force, keeping planets in orbit around stars, stars held together in galaxies and galaxies close together in clusters.

Gravity also seems to have a major role in the murkier parts of the Universe. Everything that we can see and know is made from atoms accounts for just 4 percent of the total mass of the Universe. The rest is made up of dark matter and dark energy. Gravity, the only known force that acts on this mysterious dark matter and dark energy that accounts for 70 percent of the Universe's mass, might be a sort of anti-gravity, causing the Universe to continually expand against the impacts of gravity, which would bring the cosmos's expansion to a halt.

Gravity was the first of the fundamental forces to be identified by scientists – by English mathematician Isaac Newton no less. Walking in his orchard one summer evening in the 17th century, he saw an apple fall to the ground and began to wonder if the force of attraction between fruit and Earth was the same as the one that kept the Moon in orbit. His universal law of gravitation correctly deduced that the force between two masses was proportional to their quantity of mass and that it dropped off according to the square of the distance between them.

Newton's ideas held sway until the other physicist that everyone has heard of, Albert Einstein, came up with general relativity in 1915. He reduced gravity to a problem of geometry. His idea is best explained if you imagine space represented by a flat rubber sheet. This represents space with no gravity acting anywhere. Roll a marble across the sheet and it will move in a straight line. Place a mass at the centre of the sheet, a bowling ball for example, and the rubber sheet will distort around it. Einstein argued that a star or planet did the same to space–time, the quantity physicists use to describe the arena in which gravity acts. Roll a marble across the distorted rubber sheet and it will no longer travel in a straight line but, instead, its path will curve around the bowling ball. The marble seems to be 'attracted' to the bowling ball.

General relativity likens gravity to the effects of this curved rubber sheet – stars, planets and galaxies all bend the flat sheet of space–time by different amounts depending on the mass. When they are moving relative to each other, a planet's path, for example, will be deflected around a star.

Gravity is unique among the four fundamental forces in that it has no agreed quantum mechanical description – a description that would work at the sub-atomic scale. Defining a quantum theory of gravity has been the goal of many physicists for much of the past 50 years. If it does exist, a quantum theory of gravity would work by transmitting gravitons between masses. These have yet to be found in experiments.

Unification of the forces

For as long as scientists have known about the four fundamental forces, they have looked for ways of explaining them all in a grander, all-encompassing theory of everything. Cosmologists know that, in the moments after the Big Bang, all the forces were combined, that gravity split off first, then the strong force. In the late 1970s, they had worked out that the electromagnetic and weak forces were actually manifestations of the same thing. Their next task is to incorporate the strong and gravitational forces.

The most promising unification theory, though not without its critics, is string theory, which proposes that particles are actually 11-dimensional strings that vibrate at different frequencies (see *How to know the mind of God*). Each frequency is the equivalent of a different particle, much like the different notes produced by a guitar string. The problem is, it is so difficult to test that some claim it is only as useful as philosophy. When you can't test a theory, it's hard to know if it is really science.

The Standard Model is the quantum mechanical description of the Universe as we understand it today. It contains all the known fundamental particles in the Universe.

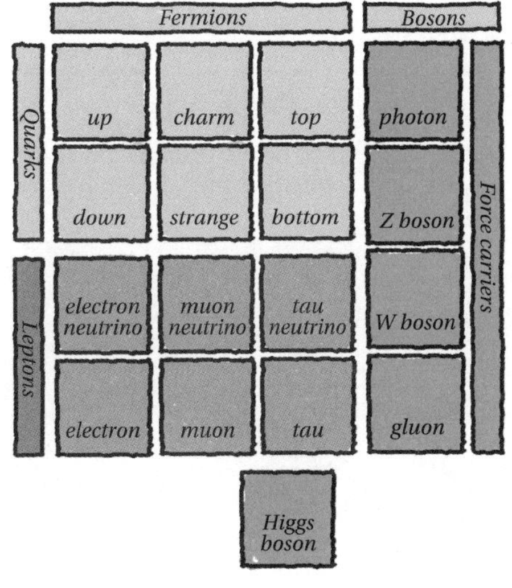

13 How to make lightning

- What is electricity?
- Force of nature
- Direct to every home
- Electrical world
- The current wars
- The future of electricity

It hardly seems necessary to convince anyone about the benefits of electricity. Unless you live very far away from the rest of human civilization (in which case I'm touched that this book is one of the things you have taken with you), you are no doubt surrounded by the direct and indirect effects of electricity, the most versatile and commonly used form of power ever harnessed by humans. And even before* Homo sapiens *walked Earth, electricity was lighting up thunderstorms and helping plants turn sunlight into sugar.

What is electricity?

It crackles, hums, can cook things, kill things or just make you tingle. Electricity is, at the same time, very ordinary and very remarkable. At its most basic, it is the result of the movement of electrical charges, in most cases electrons. If the electrons in a metal wire shudder and jump in unison, they transfer energy from one place to another – say from a battery to a motor – via an electromagnetic field, the result of a fundamental force of nature that emerged at the birth of the Universe more 13.7 billion years ago.

'I shall make electricity so cheap that only the rich can afford to burn candles.'

Thomas Edison

Well before electricity became something piped direct to every home, the effects of this fundamental force, electromagnetism, had been observed for millennia. Ancient Arabs, Egyptians, Greeks and Romans wrote about the shocks that could result from touching particular fish. Thales of Miletos, around 600 BC, had experimented with static electricity when he found that rubbing pieces of amber made it attract dust grains. The word electricity comes from the Latin for 'amber-like'. The industrialization of electricity had to wait until the 19th century when scientists worked out its theoretical underpinnings. Thanks to the experiments and mathematical calculations of British natural philosophers Michael Faraday and James Clerk Maxwell, electricity was ready to be rolled out to the world by the 20th century.

Force of nature

Electromagnetism is the force that shapes our everyday lives. It describes how positive and negative charges interact and is responsible for much more than electricity. It is the basis of light, all chemical reactions and makes things feel solid. The reason this book does not fall through the table is that trillions of negatively charged electrons on the surface of the table repel the trillions of electrons at the edges of the book. You think the book is resting on the table, but it is actually hovering above it by the tiniest imaginable distance. Electrons, locked into obeying the powerful electromagnetic force, always repel. They never touch.

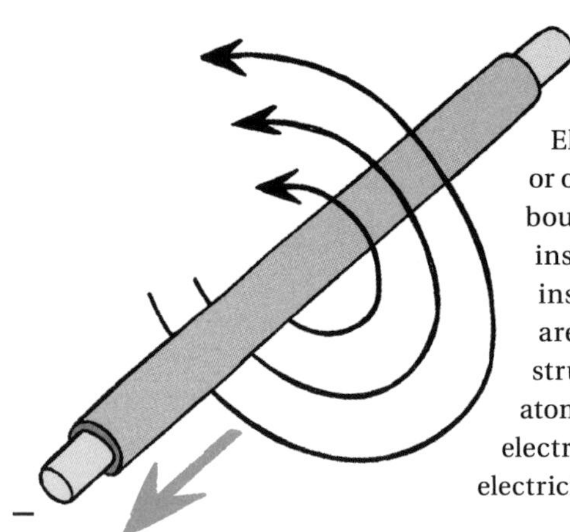

Electrons are also key to the electrical power in your home or office. Sometimes the electrons in an atom are strongly bound to the nucleus they orbit or to the molecule they sit inside. Materials like this are generally known as electrical insulators – wood, rubber, glass, air – because their electrons are not able to move around within the material. The atomic structure in a conductor, on the other hand, is such that each atom has at least one loose electron in its outer regions, an electron that can wander around in the material and conduct electricity from one place to another. Metals are all like this.

A magnetic field passes around a wire as the current flows (blue arrow) from positive to negative ends of the circuit.

The way a charged particle behaves depends on the intensity of the electromagnetic field, wherever it is in space. Switch on a kettle and the electromagnetic field around the wires and heating element spikes into overdrive, feeding energy to the electrons in the metal and causing them to shake. That shaking knocks around the water molecules in the kettle so much that they also gain energy and start to zoom around more quickly, eventually breaking free from the liquid – causing the water to boil.

Direct to every home

With an increasingly digital world dependent on electrical power for a growing number of devices, electricity is going to remain the primary way we get our energy at home or in the office. But how does that power reach the sockets? And how is it made in the first place? To understand that, we need to go back to James Clerk Maxwell. He found that electricity and magnetism were two aspects of the same thing: wherever you found an electrical field, there was also a magnetic field at right angles to it. If you made one, you created the other too. An electrical generator uses this principle to convert mechanical energy into electrical energy. Move a large magnet around a coil of metal wire and the electrons in that wire will start to move because of the changing magnetic field they experience. Moving electrons, you will remember, is the start of an electrical current. Shifting the magnets around in the first place requires some sort of external source of power – this could be anything from a hand-crank to a windmill. Modern power stations use steam turbines to turn the generator magnets. The turbines are fed by water that has been heated by burning a fuel (such as wood, coal, oil or natural gas), or the energy released when atoms split in a nuclear reactor.

The strength of an electrical current can be measured by the number of electrons moving along a wire at any given time. When six quintillion electrons – that is 6 followed by 18 zeros – rush past every second, we measure one 'amp' of current flowing (a unit named after the French scientist André-Marie Ampère). The Italian scientist Alessandro Volta gave his name to another important unit of electricity, the 'volt'. Think of this as a measure of the force behind the electrons, the pressure that makes them flow in the first place.

Electrical world

In an electrical circuit, you connect a power supply (which has one positive terminal and one negative terminal) to an appliance. Electrons come out of the negative terminal and all they want to do is balance out the electrical charge around the wires by travelling to the positive terminal of the power supply. How much charge difference there is between the positive and negative terminals determines how much voltage (or pressure) there is for the electrons to move.

As far as the electrons are concerned, anything you put in the path between them and the positive terminal, called the load, is not important. The electromagnetic force compels them to do anything they can to transfer their energy to their destination, even if that means passing it through a light bulb or turning a wheel along the way. Our global use of electricity is entirely based around the electrons' dash for the positive terminal. In an old-fashioned light bulb, the electrons smash into the metal atoms in the filament, transferring some of their energy and making the atoms vibrate more quickly: the result is heat and light. How hard electrons find it to fight through the circuit is measured by the resistance of a component (measured in ohms). The more ohms, the more of the electron's energy is wasted as heat, and the more of it is used up trying to battle through the material. In an electric motor, the magnetic field created by the movement of electrons creates movement in other magnets (via north–south repulsion and attraction) arranged around the wire. Keep the electricity flowing and the magnets (and therefore the wheel attached to them) keep turning around the wire.

Each electrical appliance uses the moving electrons in its own way and how much energy each circuit uses up can be measured thanks to a unit named after Scottish engineer James Watt. Multiply the rate of electrons passing through the circuit (the amps) by the amount of 'pressure' they rush by with (volts) and you get the energy used per second by an appliance. A 100-watt light bulb, for example, uses 100 joules of energy per second. And, if it is an old-fashioned incandescent bulb, around 95 joules of that is being wasted as heat every second – a good reason to change to compact fluorescent bulbs, which are at least five times more efficient at converting the electricity into useful light.

The current wars

Not unusually for an industrial technology that became fundamental and widespread, the early years of electricity were characterized by huge disagreement. US inventor Thomas Edison and Serbian-born Nikola Tesla spent the last few years of the 19th century battling to convince the world that their way of generating and distributing electricity was better and more efficient than the other. It was a bitter contest because, by then, the rivals knew that the winner would decide how electrical power systems would be built across the world. Their disagreement rested on the fact that electricity comes in two basic flavours: direct and alternating. Edison advocated the former, while Tesla was a fan of the latter. Direct current (DC) is the type where electrons flow from negative to positive terminals, the type produced by batteries and solar panels. Alternating current (AC) is the same except that its direction changes many times every second, as its name suggests. Edison had a lot of money invested in DC at the end of the 19th century and tried to discredit Tesla's AC idea by insisting it was too dangerous (he electrocuted an elephant to death to prove his point, fortunately the only casualty of the current wars).

Edison's public relations attempts did not, however, succeed in the end. In the UK and most of Europe, the power coming out of your sockets switches direction 50 times per second, while in the US, it changes 60 times per second. AC is the electrical transmission method of choice around the world. Both AC and DC have pros and cons but Tesla won mainly because, with AC, it is easier to step up the voltage of your electrical current using a simple device called a transformer. Changing voltages makes it easier to transport the electricity over long distances in electrical grids without wasting too much along the way in the wires.

Electricity can only flow around a complete circuit. A switch turns the current on or off by completing or breaking the circuit. The switch in this circuit is in the open, or 'off' position, and no current can flow.

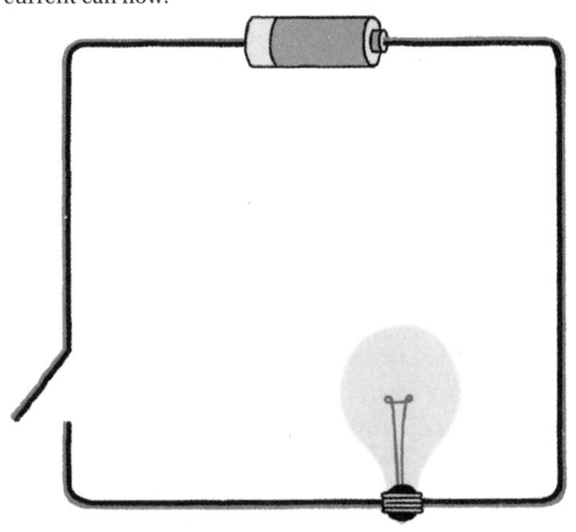

Power grids tend to conduct electricity over long distances at several hundreds of thousand of volts, which reduces the amps of the current (and therefore the thickness of the wire needed to successfully carry it). So many volts can be dangerous if anyone got close, however, so when the electricity gets close to its destination, the voltage is stepped down, again using a transformer. A plug socket in your home or office delivers power at a few hundred volts, even though the power station it came from might be generating it at several thousand volts.

The future of electricity

Electricity underpins our modern world. By understanding and manipulating atoms, we have created a power source that is predictable, reliable and secure. Unfortunately, the rapid electrification of the world has also contributed to one of our gravest problems – power stations belching out carbon dioxide over the past century threaten to warm our planet irrevocably, changing habitats and placing millions of people at risk of floods and famine. Finding affordable sources of clean electricity is one of the biggest goals for the 21st century. To rid the world of large-scale use of fossil fuels will be an enormous task. But there are good ideas out there, based around the energy flowing around the world every day. Engineers have the technology to harness the power of the Sun, wind, waves and even the heat inside Earth to turn turbines and generate electricity. The question is whether anyone will invest fast enough to build the amounts we need to replace fossil fuels and keep pace with the growing need for electricity around the world at the same time.

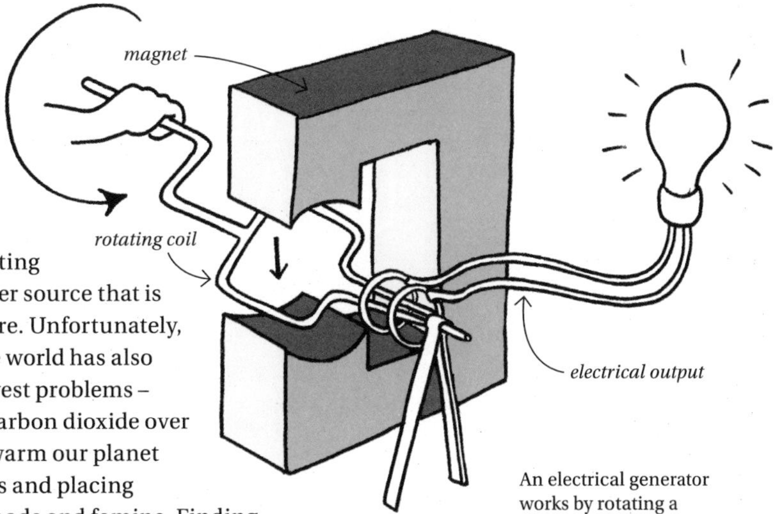

An electrical generator works by rotating a coiled wire inside a magnetic field. This creates an electrical current in the wire, which can be used in a circuit.

Perhaps the ghost of Edison can help. One of the most ambitious plans to replace fossil fuels with clean electricity has been proposed by various government agencies in Europe, and it relies on one of the cast-off ideas in the history of electricity. The Desertec consortium thinks that building vast solar-panel arrays in places such as the Sahara desert, the outback in Australia or the Arizona desert could provide more than enough electricity for our growing world. Once generated, it could be transferred via long-distance cables to wherever it is needed, thousands of kilometres away. Normal AC networks would be no good for this task since the energy losses over such a long distance would be huge. Instead, Desertec engineers propose using high-voltage DC cables – which are more expensive to build but far more efficient to use over long distances.

Whatever the future of electricity looks like, Desertec or a scheme like it will, no doubt, play a part. And that future will only work because of a concept resurrected from the history of electricity. Edison lost the current wars but he would probably raise a smile that his idea might go on to help save the world anyway.

14 How to put the Universe to work

- Energy makes the world go round
- The motion of small things
- The motion of collections of things
- How it applies to the real world
- The age of steam

The steam trains that survive today are a reminder of the era of mechanization, which eventually spread across the world, bringing the Industrial Revolution and the precursors of our modern society. The engineers who designed the first steam engines were looking for a better way to move heavy things and increase production in ever-bigger factories. In the process, they ended up uncovering fundamental truths: the humble steam engine, with its dirt, metal and oil, not only changed the world, but changed how we understood that world.

Energy makes the world go round

Look at things long enough and there are lots of things about the Universe that are strange or unsettling. The inherent uncertainty of quantum mechanics, for example, or the idea that how fast time ticks by depends on how fast you're moving (see *How to age slower than your twin*). Science is an exploration of the natural world, and exploration is unsettling by its very nature. But how do you know if your new idea, the one that overturns centuries of previous ones, is still an apt description of that natural world? Fortunately, there is a handle you can always grab onto, something that will always tell you whether your idea is part of reality. And it has to do with energy.

'In this house, we obey the laws of thermo-dynamics!'

HOMER SIMPSON

It has been true during the entire history of the Universe, from the emergence of the first forces, the creation and annihilation of particles and stars, to the creation of Earth and all its life forms. It has worked through the highs and lows of human civilization, in battles, in the building and destruction of cities. It is true in the way our bodies move and in the way cars burn fuel. Carve it over your front door for there is nothing more true: you cannot create or destroy energy. It is true for us, our planet, our galaxy and every object and movement in the entire Universe. There is a fixed amount of energy in existence, which can change from one type to another (chemical energy in food turns into kinetic energy in your arm as you lift up a cup), but the amount of energy will always be the same. If you come up with a new scientific idea, check how it affects this basic idea – if it seems to create or destroy energy, think again.

The conservation of energy is fundamental to our understanding of nature. It works in the weirdest situations in quantum theory and it has been used to predict the orbits of stars and planets. The only time it might look shaky is inside the hearts of stars, where energy seems to spew forth from nothing. This was solved, however, by Albert Einstein and his equation, $E=mc^2$. Whenever new energy seems to be created, a corresponding (though small) amount of matter disappears. And so the balance is maintained.

The motion of small things

Extracting useful work from the Universe requires an understanding of how you convert one form of energy into another. Heat is a very common form of energy but it is unfocused and useless for doing anything productive (save, of course, keeping something warm). If you want to move something from here to there, heat won't do it, but it is not a bad place to start. Heat is the macroscopic result of the random movement and vibration of molecules. Imagine you have a box filled with gas. Its temperature, pressure and volume are all related and can be accounted for by the movement of the molecules of that gas. The temperature of the gas, for example, is a measurement of how fast its molecules are moving. Add heat to the box and each of the gas molecules gains some energy and is able to move more quickly. This has the effect of making the gas hit the sides of the box with more force, thus increasing its pressure. That happens if you want to keep the volume the same.

The aeolipile (also known as Hero's engine) is a rocket-style engine that spins when heated, as steam is fired out of the nozzles. Invented in the first century AD, it is thought to be the first recorded steam engine.

If, instead, you want to maintain constant pressure or temperature with the addition of heat, you could change the other variables. To maintain constant pressure while the gas is heated, let it expand as the temperature rises. Want to maintain constant temperature while heating your container? Reduce the volume as the pressure increases.

This behaviour was described after experiments carried out in the 17th and 18th centuries. The work of scientists including Englishman Robert Boyle, Frenchmen Jacques Charles and Joseph Louis Gay-Lussac, and Italian Amedeo Avogadro led to a combined 'ideal gas' law, which shows how the product of volume and pressure is proportional to temperature. Ideal gases are so called because the law assumes that the molecules behave like floating billiard balls that bounce off each other and don't react with each other.

The motion of collections of things

The ideal gas law describes the relationship between heat and kinetic energy in gas molecules. As such, it became part of the staging ground for the development of thermodynamics, which is a more general description of energy and how small-scale molecular movements translate into large-scale properties of materials.

The beginnings of thermodynamics are generally credited to the French scientist Sadi Carnot. The father of thermodynamics published *Reflections on the Motive Power of Fire* in 1824, in which he described how energy flowed through engines and how heat and useful power were related. It laid down the scientific principles behind the steam engine and is still in use today – Carnot's basic descriptions of how much useful energy it is possible to extract from a source of heat is relevant even for engineers working on the most cutting-edge jet engines.

We're already familiar with the first law of thermodynamics, and will look at it in further detail below. Unusually for a set of laws of nature the ideas start before the first law, with a concept that was thought so trivial that it was originally left out. If two thermodynamic systems (such as a box of gas, or anything that can be isolated and studied in terms of its energy) are in thermal equilibrium with a third system, then the two systems are also in equilibrium with each other. This means that if system A and system B have the same amount of heat energy, and system B and system C have the same amount of heat energy, then system A and system C must also be the same.

This 'zeroth' law makes it scientifically possible to make a thermometer, for example, to measure whether two objects are at the same temperature, rather than needing to bring the objects together and watching if any energy flows between them. Even though this law was added last to the study of the subject, the zeroth law was thought to be more fundamental than the others, hence its ever-so-slightly bizarre designation.

Starting at the original beginning, though, is the conservation of energy. The first law of thermodynamics, you will recall, says that energy cannot be created or destroyed, merely changed from one form to another. The second law is the mathematical answer to why you never get something for nothing. It ushers in a new variable in the properties of a thermodynamic system: entropy, a measure of the amount of order. Order can, for example, be measured as the arrangement of molecules – the water molecules in a tray of ice cubes are more ordered than the same molecules in a pan of boiling water. In the first scenario, the molecules sit in well-defined places and vibrate around fixed points. In the second, the molecules have much more energy and can wander freely and unpredictably throughout the fluid. Scientists would say that the entropy of the first scenario is lower than the entropy of the second.

Entropy is the measure of how ordered a system is. In an ice cube, the water molecules are precisely arranged, so the entropy is low. In a puddle of water, the molecules are moving randomly, so entropy is higher.

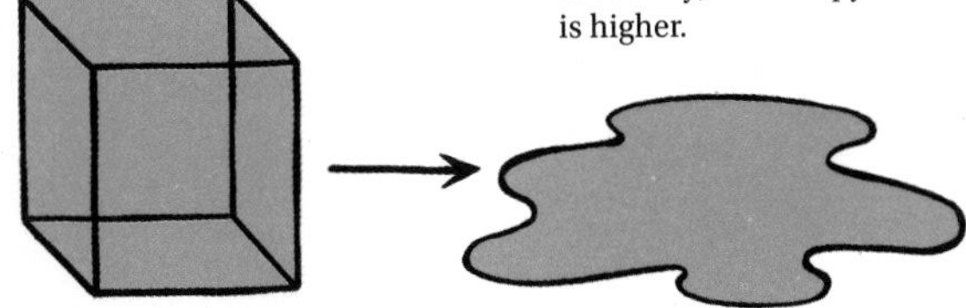

Similarly, a cup has more entropy when it is in pieces on the floor than when it is sitting in one piece on the kitchen counter. With that in mind, the second law – that the entropy of an isolated thermodynamic system can never decrease – has several implications. Chief among them is that heat cannot spontaneously flow from a cold object (low entropy) to a hot object (high entropy). More intriguing, the second law finally introduces the idea of a fixed direction for time's arrow – how time can flow only in one direction.

Every other law of physics works the same whether time is going forwards or backwards. But, however long you leave it, a boiling pan of water is unlikely to become a tray of ice cubes. A smashed cup will never reassemble itself. Entropy never decreases if you leave a system alone. Water only freezes into ice cubes, for example, if you use a refrigerator to freeze it and that requires energy, some of which is wasted as heat. The entropy of the water goes down but the wasted heat ends up raising the overall entropy of the Universe.

The third, and final, law of thermodynamics defines the lower limits of temperature and entropy. If you took all of the thermal energy out of a system (all of the energy related to movement, which is everything that is not mass), its temperature would reach absolute zero and its entropy would be zero. At absolute zero (defined as 0 K and equivalent to –273.15°C or –459.67°F), all molecular processes stop.

How it applies to the real world

The energy laws apply across a huge number of situations. A thermodynamic system could be the air in a tropical cyclone, or a single particle in an atom. It could also be, as Carnot originally defined it, the air or steam in a steam engine. In a typical engine, thermodynamics can be used to determine how heat flows around and how it can be used to do something useful – drive pistons, for example, or turn an axle. A steam engine (or, for that matter, any heat-based engine) works because the zeroth law states how heat naturally flows from something hot to something colder. These two bits in an engine are usually called the hot and cold reservoirs. As energy flows between the two parts, useful energy (called work) can be extracted out of the flow. This useful work could mean using heat to expand gas in a cylinder, say, which then pushes on a piston and does something useful. The gas can then be compressed and the cycle started again with more heat. Carnot showed that you could predict the efficiency of any such engine by comparing the temperatures of the hot and cold reservoirs (for a steam engine, the steam itself and the air around it). Whatever you might use to transfer heat around in your engine, Carnot's rule is the same and the efficiency is determined purely by the temperature difference between the two reservoirs.

The need for a cold reservoir shows that no engine can be 100 percent efficient. You always need to dump some of the energy into a cold reservoir, otherwise there will be no flow in the first place. To work as efficiently as possible, a steam engine should heat water to the highest possible temperature and release the waste heat at the lowest possible temperature. In practice, this means that a traditional steam engine has a maximum efficiency of around 10 percent, though this can be improved to 25 percent with technologies that condense the exhaust steam and use waste heat to warm the incoming water. Modern engines are better but still nowhere near 100 percent efficient – a diesel engine is around 50 percent efficient at converting chemical energy in its fuel to useful energy that can move a vehicle. Petrol engines are much less efficient than that.

The age of steam

Hero of Alexandria designed and built a prototype steam-powered device in the first century AD. The 'aeolipile' was a brass sphere suspended with two nozzles mounted onto it, facing in opposite directions. Steam rushed out of the nozzles and forced the sphere to spin at impressive speeds. As steam engines go, the aeolipile proved a point but it was nothing more than a toy. It took almost two millennia for English blacksmith Thomas Newcomen to build something that approached what we would think of as an engine. He used it in the 18th century to operate a pump that could remove water from mines.

James Watt improved Newcomen's design by reducing the amount of coal it needed and modifying it to work with factory equipment. By the start of the 19th century, Richard Trevithick had started using high-pressure steam in miniaturized engines so that they could be used in transport. Steam engines became more efficient and more powerful but were gradually replaced in the 20th century by electric motors and fossil-fuel-powered internal combustion engines in road vehicles. The principles behind the steam engine are far from forgotten, however: they are at the heart of virtually all electricity power stations (see previous chapter). Advanced steam-powered turbines use water heated by fossil fuels, while nuclear reactions turn giant electrical generators to power the world, our homes and offices, our mobile phones and even electric trains. The age of steam is not over yet.

A steam engine uses steam at different pressures to move a piston backwards and forwards, creating useful motion for trains or other industrial applications.

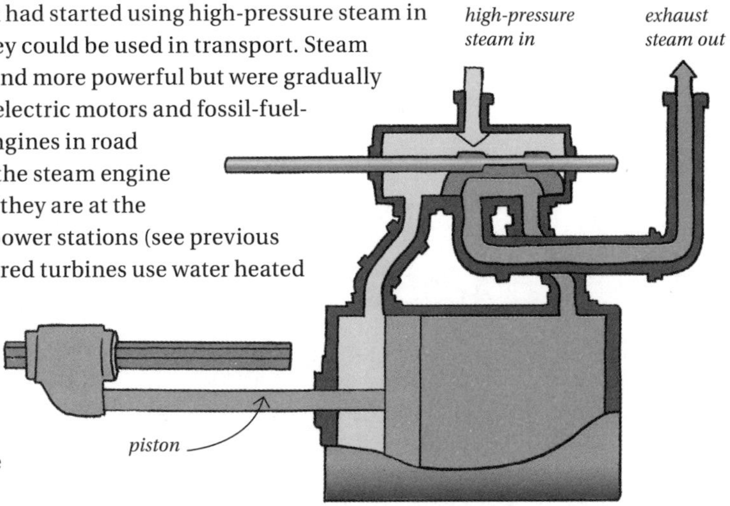

15 How to split an atom

- The mechanics of nuclear physics
- Discovery of the nucleus
- How to make a nuclear bomb
- Nuclear power
- Can splitting atoms save the world?
- The next step: only 50 years away
- How fusion works
- Do we have the technology?

At 8:15 am on August 6, 1945, a mushroom-shaped cloud appeared over the Japanese port city of Hiroshima. Unleashed by an American-made bomb called Little Boy, it became the unimaginably violent sign that the world had moved into a new age. It was a mixed time for scientists. In less than 40 years, they had gone from having only the vaguest idea of what an atom was, to building a detailed model that included the nucleus, electrons, and a theoretical understanding of how to manipulate them all – with potentially deadly results.

The mechanics of nuclear physics

The phrase 'splitting an atom' has entered the vernacular as one of those things scientists do, shorthand for the sophistication and power of modern technology, but without much description of what the process actually involves. Most of an atom is empty space and virtually all of its mass is bound up in a tiny nucleus at its centre, composed of protons and neutrons. Splitting an atom actually means breaking apart this nucleus, which is bound together with two of the strongest forces known to exist in the Universe. These forces are so strong that the only atoms scientists have so far been able to split successfully are those that are already falling apart naturally, atoms that have huge nuclei with scores of protons and neutrons. These unstable atoms can be nudged along to splitting by firing neutrons at them. When the nucleus splits, it results in two new atoms and the release of some of the energy bound up in the forces between the proton and neutrons.

Discovery of the nucleus

Our knowledge of the building blocks of matter – atoms – is a science that is barely a century old. In 1900, it was assumed that these building blocks were made of a positively-charged lump of mass studded with negatively-charged electrons – the so-called 'plum pudding' model. In 1907, that model came under attack. Anglo-New Zealand physicist Ernest Rutherford fired alpha particles, sourced from a radioactive element, at a thin leaf of gold foil. Most of the particles passed right through the foil, which did not fit with the idea that atoms were solid lumps of positive charge. If that were true, no alpha particles should have passed through the gold, however thin the foil. Most of the gold, it seemed to Rutherford, was empty space.

Knocking down the plum pudding model was already a remarkable feat. But Rutherford went further. A very small number of his alpha particles did not pass through the foil, but instead bounced right back towards the source. Rutherford was taken aback by this result and would later write that it was as

'In the current fashionable denigration of technology, it is easy to forget that nuclear fission is a natural process. If something as intricate as life can assemble by accident, we need not marvel at the fission reactor, a relatively simple contraption, doing likewise.'

JAMES LOVELOCK

remarkable as firing a 15in (38cm) shell at a piece of tissue paper and having it come back and hit you. What was causing this strange rebound effect? Rutherford had unknowingly found the atomic nucleus.

How to make a nuclear bomb

Knowing that the nucleus was the centre of the atom and that energy and mass were equivalent – Einstein's great insight, epitomized by the most famous equation in physics, $E=mc^2$ – were two of the greatest scientific discoveries of the 20th century. But together they also led directly to the 20th century's most dubious invention: the nuclear bomb. The bombs used by the US military over Hiroshima and Nagasaki were made in the 1940s at a top-secret facility at Los Alamos National Laboratory, under the umbrella of the Manhattan Project. The weapons they designed were rudimentary versions of modern nuclear warheads, and the underlying principles behind them are identical.

To start with, you need atoms that will split when given a nudge. The most commonly used 'fissile' material is uranium, a naturally occurring heavy metal that comes in two types: isotopes called uranium 238 and uranium 235. Given time, both isotopes are radioactive and will decay into other elements. But only uranium 235 can be forcibly split when neutrons are fired at its nuclei. This is the key ingredient for a nuclear bomb. When a uranium nucleus breaks apart, it gives out energy and more neutrons, which go on to split other uranium nuclei. Get enough atoms splitting and you start a chain reaction. The tiny amounts of energy released as each nucleus splits build up over billions of atoms to create that immense and unmistakable explosion.

Fortunately for the sake of world peace, however, nuclear explosions are much easier said than done. Natural uranium overwhelmingly consists of the 238 isotope, which bounces back any neutrons striking its nucleus and is generally quite useless for bomb-making. For the purposes of a bomb, natural uranium needs to be processed in order to concentrate the 235 isotope. This is harder than it sounds: for every 25,000 tonnes of uranium ore, only 50 tonnes of metal are produced and less than 1 percent of that is uranium 235. And no standard extraction method will separate the two isotopes because they are chemically identical. Instead, uranium is reacted with fluorine, heated until it becomes a gas and then decanted through several thousand fine porous barriers. This separates the uranium into two types: 'enriched', containing mostly uranium 235, and 'depleted', containing mostly uranium 238. For nuclear bombs, uranium needs to be enriched so that it is around 80 or 90 percent uranium 235. Get around 50 kg (110 lb) of enriched uranium – the critical mass – and you have a bomb. Any less and the chain reaction would fizzle out before it could create a substantial explosion.

A nuclear chain reaction occurs when the splitting of nuclei becomes self-sustaining. If a uranium-235 nucleus is hit by a neutron, it splits into two and releases more neutrons. These go on to split further U-235 nuclei (though U-238 nuclei are unaffected) and release even more neutrons, which go on to split further nuclei.

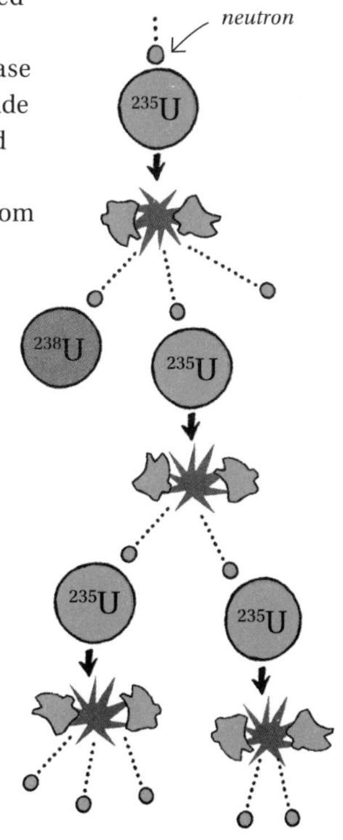

Nuclear power

Atomic weapons are one of the most devastating inventions of the 20th century. But splitting atoms has also been used for peaceful purposes: chances are that some of the electricity powering your lamp while you read this book came from a slow-motion version of an atomic bomb, a nuclear reactor. The world's first large-scale nuclear power station, Calder Hall, was opened in Cumbria, England, in 1956. It kept going until 2003 and produced electricity in much the same way as a coal, oil or gas station: heating water into steam, which is then used to turn turbines to make electricity. In the case of nuclear plants, the heat comes from the fission reactions in fuel rods made from uranium. The fuel rods are immersed in a pressurized water tank and bombarded with high-speed neutrons. The energy released whenever a nucleus falls apart is used to heat the water and the reactor is prevented from blowing up by using control rods, usually made from graphite, that can absorb excess neutrons. These rods can be inserted into the reactor to different depths to control the speed of the reaction.

Can splitting atoms save the world?

A century after the discoveries of Rutherford and Einstein and decades into becoming the poster subject for anyone worried about our to destroy the planet, nuclear physics could emerge as the world's saviour from the biggest threat we have ever faced. Thanks to its lack of carbon dioxide emissions, nuclear power is fast becoming one of the suite of suggested solutions to tackle climate change. The UK recently reaffirmed targets to build more plants, to the dismay of many environmentalists who legitimately cite the cost and huge question-marks about what to do with radioactive waste, some of which will remain toxic for thousands of years. But will the nuclear power stations of the future be as bad as those of the past? Many decades of development means that the next generation of nuclear power plants are a world away from the ageing stations of today.

One of the biggest changes between 'legacy' nuclear power and more modern plants, such as those in South Korea and China, are the safety systems. Instead of engineered safety mechanisms, where valves and pumps bring cooling fluids, for example, into the reactor core in the event of an accident, modern designs rely on passive systems. These have fewer moving parts and require less maintenance. According to Westinghouse, which designs nuclear power stations, their most advanced designs – such as the AP1000 – are around 100 times safer than existing stations.

The emergency cooling water in an AP1000, for example, is above the reactor core. In the event of an accident, the water just falls on to the core. As it begins the cooling process, it converts to steam, hits a stainless steel barrier at the top, condenses to water and rains back on the core. The Westinghouse design also uses uranium 60 times more efficiently, producing 10 percent of the waste of nuclear power stations today.

The next step: only 50 years away

For all their improved efficiency and safety, the next generation of nuclear power stations will still split nuclei to create electricity. Atoms, though, don't have to be torn apart to release energy. Nature prefers the opposite method of releasing energy – fusion, the process by which stars are powered. In theory, fusion power is cheap and clean. It produces no greenhouse gases and there is a plentiful supply of fuel, in the form of seawater. Fusion has kept physicists and engineers busy for more than 50 years and many of them believe that such a power source will be a major step in solving our increasing demands for energy. The work has had its share of sceptics too, who point out that trying to reproduce a star on Earth is too difficult, dangerous and expensive. To test who is right, a coalition of nations are working on the International Thermonuclear Experimental Reactor (ITER) project, which aims to build the first prototype fusion power plant at Cadarache in France. ITER began life in 1985, when Ronald Reagan and Mikhail Gorbachev called for scientists to prove that fusion was a scientifically and economically viable way of producing electricity. Europe and Japan soon joined Russia and the US as partners and now the multi-billion dollar collaboration also includes China, Canada and South Korea.

How fusion works

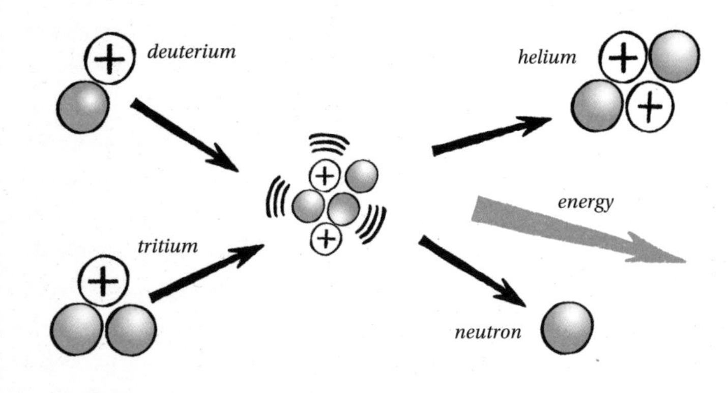

At the heart of each star there are countless billions of hydrogen nuclei, which are single protons. They fuse to form helium nuclei (two protons and two neutrons), plus energy. It doesn't happen easily, however. Despite the immense gravitational pressure and temperature at the core of every star, it takes millions of years to fuse two nuclei together, such is the repulsive force between two protons. On Earth, generating energy using a reaction that takes so long would be next to useless. So, instead of hydrogen, physicists fuse two of its isotopes – deuterium and tritium – which are heavier than hydrogen and can be made to fuse more easily.

In this nuclear fusion reaction, deuterium and tritium (both types of hydrogen), fuse together to form an atom of helium, releasing much more energy than in nuclear fission.

Deuterium is abundant in sea water. Tritium is harder to come by and has to be made inside a fusion reactor. Even so, we have enough resources to last several million years. The fuel is placed inside a donut-shaped chamber, called a torus, at the centre of the machine and heated to create plasma at 100 million °C. The deuterium and tritium fuse to form helium, energy and spare neutrons, which are absorbed by a lithium shield around the torus. When the neutrons hit the metal, more tritium is produced, and this is fed back into the torus.

Do we have the technology?

The principles behind fusion were established in the middle of the 20th century, but recreating the process that makes stars shine was way beyond the technology available at the time. Scientists reckoned early on that very strong magnetic fields would be needed to contain the fuel in a reactor but, back in the 1950s, they had no reliable way of producing them. Fast-forward 50 years and improved computer simulations of how the fuel would behave in the reactor and test experiments in places such as the Joint European Torus in England have proved that fusion can work.

But no-one has yet managed to release more energy from a fusion reaction than was put in to start the reactions in the first place. ITER aims to put this right and produce 500 megawatts of power, 10 times its predicted input. If ITER works, the implications for energy and the planet's environment will be profound. Perhaps, in its second century, the nuclear age will become the poster subject for hope rather than fear.

16 How to know the mind of God

- The fix – a mythical theory of everything
- Beyond the Standard Model
- String theory
- What about mass?
- How do we test any of this?

The two pillars of modern physics – Einstein's general theory of relativity and quantum mechanics, developed by Niels Bohr, Werner Heisenberg and Paul Dirac – don't quite agree with each other. And that is a big problem if we ever want to answer some of the biggest questions in physics, such as what actually happened in the Big Bang and where the Universe came from. Do you use the equations of general relativity because there is an enormous amount of mass? Or do you use quantum mechanics because it's all in such a small space?

The fix – a mythical theory of everything

Albert Einstein's general theory of relativity described how the gravity around the very largest objects warped the space around them, turning Isaac Newton's elegant force of gravity into an exercise in geometry. Quantum mechanics showed that the world at subatomic level was unpredictable, that the behaviour of the elementary particles lay in the hands of chance. Bridging the chasm with a so-called 'theory of everything' has consumed theoretical physicists for decades. Einstein himself spent the last decades of his life trying to unify general relativity and electromagnetism – he had no inkling of the strong and weak forces lurking in the nucleus, and did not care much for quantum theory. He had got nowhere by the time he died in 1955.

Others toyed with the idea of a quantum theory of gravity but most of the work of physics in the past century was spent in developing and understanding quantum mechanics itself. The rock-star physicists of the 1950s and 1960s – including US greats Richard Feynman, Murray Gell-Mann and Steven Weinberg – busied themselves in the work of characterizing and turning quantum mechanics into the robust theory we know and use today (it is the basis of all electronics, for example). They won several Nobel prizes on the way to building and fine-tuning the Standard Model of particle physics, a precise quantum mechanical description of all the subatomic particles that are known to exist. None of this work, however, moved quantum mechanics closer to Einstein's general relativity. Unification work never went away, though. The past few decades have seen much activity, with theorists laying the groundwork for potential ideas to unify the four forces of nature (see *How to join up the Universe*), helping to answer yet more conundrums, such as where does mass come from? And why is most of the Universe missing? There have been no major discoveries in particle physics since the Standard Model 25 years ago. With the start of experiments at the Large Hadron Collider (LHC) in Geneva, however, a new generation of scientists are getting ready to take this carefully built piece of physics history, and smash it to pieces.

'I want to know how God created this world. I'm not interested in this or that phenomenon, in the spectrum of this or that element. I want to know His thoughts; the rest are details.'

Albert Einstein

Beyond the Standard Model

The Standard Model describes the properties of 12 matter particles called fermions and the various force particles that define how these fermions interact with each other. The matter particles are made of six quarks (some of which make up protons and neutrons, while others are so heavy they only survive for fractions of seconds before decaying into lighter particles) and six leptons (including the electron and neutrinos). The force particles (called bosons) include the particle of light – the photon – and gluons that stick quarks together in the nucleus of an atom. There is also antimatter, made up of particles that are identical to normal matter but with opposite charges: the positron, for example, has the same mass as an electron but is positive; an anti-proton is made up of anti-quarks and has an overall negative charge.

If the Standard Model could be extended to include aspects of gravity, it would be a major step in unifying the forces. The graviton has been proposed as the force-carrying particle for this force but, so far, has eluded experimental observations. This hasn't stopped physicists from proposing extensions to solve one of the big headaches – the fact that most of the Universe is missing. Around a quarter of the Universe is thought to be made of 'dark matter' that does not radiate and cannot be detected. Some theorists hope that an idea called 'supersymmetry' might help. This extension to the Standard Model proposes that every known particle has a heavier, super-symmetric twin. The electron is twinned with an as-yet-undiscovered selectron, quarks pair up with squarks, and photons and neutrinos find that they have photinos and neutralinos. Fortunately, the last of these particles (also the lightest) has some of the properties predicted for the mysterious galactic dark matter.

Another replacement for the Standard Model goes by the name 'technicolour'. While the notion of supersymmetry takes the same familiar forces of nature and adds extra particles, technicolour invokes a bizarre fundamental force that we have not yet detected. It says, for example, that quarks are not single particles, but a complex of smaller particles called techniquarks.

String theory

Supersymmetry is also an essential part of the one theory in physics that is closer than any of the others to becoming the theory of everything (though not without some controversy). String theory can explain quantum theory and relativity, and answer most of the unification problems. But it requires the world to have 11 dimensions and its ideas have never been tested experimentally. In fact, it operates in an arena so tiny that it may never be testable. Which raises the question of whether it is a scientific idea at all.

The origins of this arcane and fiendishly complicated mathematical world came in the late 1960s as a way to describe the strong nuclear interaction, the force that stops the nuclei of atoms from flying apart due to the repulsive force between protons. Protons are made of particles called quarks, which are held together by yet more particles, called gluons, that transmit the strong nuclear force. The initial string theories emerge as a way to answer the question of why quarks and gluons were never seen by themselves, even when atoms were smashed apart in particle accelerators. In simple terms, string theory said that perhaps gluons and quarks were at the ends of a string of energy and, since you can't have a single-ended string, you can't have individual quarks and gluons. The idea fell out of favour after physicists came up with a more robust way of describing the strong nuclear force with quantum chromodynamics, a precise quantum mechanical description of the interaction between quarks and gluons.

But string theory did not die altogether. It later emerged as a way to explain not just the strong force, but – perhaps – all the four fundamental forces. The earliest versions of this incarnation of string theory posited that everything in the Universe, from Earth to the book you are holding now, and all the forces that are acting upon them, is made of tiny vibrating strands of energy, the so-called strings. For a while there were numerous competing string theories and it took US scientist Ed Witten, the de facto chief of string theory and the closest thing physicists have to a modern version of Albert Einstein, to bring order to the chaos. With M-theory, he showed that all the competing theories were, in fact, different facets of the same idea.

String theory says that, at the most elementary level, everything is made from vibrating strings. These can be open (like the ring on the diagram below) or closed (the squiggles) or even multi-dimensional membranes (the blue panel) stretching across the entire Universe.

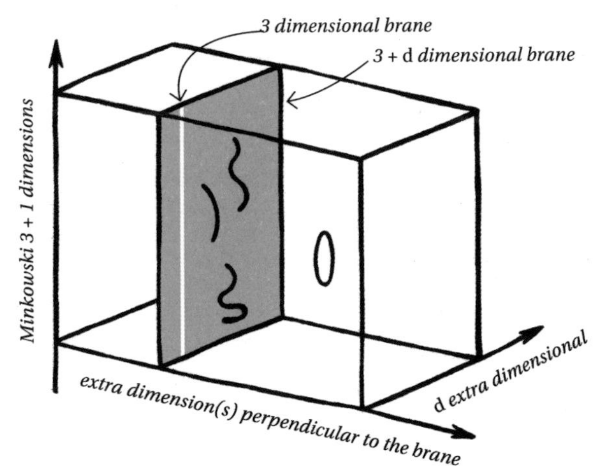

Before M-theory, strings existed in a world with 10 dimensions, including a dimension of time, the three dimensions of space and six extra dimensions, curled up so small that they are invisible. Witten's M-theory also needed an extra dimension for space, taking the total to 11, and suggested that this world contained not only strings, but objects that looked more like surfaces or membranes. These 'branes' could exist in three or more dimensions and, with enough energy, could grow to huge sizes – even as large as the whole Universe. Extra dimensions could also tackle the problem of why gravity is so weak. Perhaps it actually exists as a strong force in an as-yet-undiscovered dimension. We experience it as weak because only part of its strength trickles into our world.

According to Einstein's general theory of relativity, the mass of a star bends space-time around it so that the path of anything that passes nearby by is curved around the star. We describe that effect as the force of gravity.

A lot of mathematical work has been done by some of the best scientific minds in the world to develop the ideas of string theory. It is by far the most popular idea for the unification of quantum mechanics and gravity. But the idea goes so far beyond our physical experience of the world that many critics have lined up to argue that it should be considered more as a work of philosophy than the ultimate scientific description of nature. And, until experiments can prove otherwise, those arguments will continue.

What about mass?

It seems odd to question where mass came from. It is just there, right? If you want to get more specific, perhaps you could say mass is the cumulative amount of the fermions – the quarks and leptons – in an object. Add these tiny masses up and you have a total mass for an atom or bigger object. If only it were that simple. By the earlier logic, you might think that this book's mass is the sum of all the atoms in the sheets of paper it contains. Dig deeper and you will find something very strange, though. It's almost as if the matter particles, the quarks and leptons, have no mass of their own.

Physicists never really understood why this happens until a shy physicist from Edinburgh came up with a possible solution. Peter Higgs reckoned that the building blocks of matter were weightless when they were formed in the first few minutes after the Big Bang, but that something switched on shortly afterwards to make them gain mass. That something was an as-yet-undiscovered field permeating all of space, which clings to particles wherever they are and gives them a drag that we interpret as mass. Different particles interact more strongly than others with the Higgs field. Particles of light are oblivious to it and therefore are massless; heavier particles move through the field like someone wading in mud. To feel the effects of this field, his calculations showed that fermions must interact with the field via an undiscovered particle, the Higgs boson, which bestowed mass upon them.

How do we test any of this?

Buried deep under the Jura mountains near Geneva, straddling the border of France and Switzerland, is a circular tunnel, 27 km (16 miles) in circumference. It has already been the site of some of the most important scientific experiments in history and, very soon, will continue that tradition into the 21st century. This is the location of Cern's much-heralded LHC, a particle accelerator that will smash protons together at almost the speed of light. The resulting explosion will tear the particles apart and, momentarily, create conditions that have not been seen in the Universe since the seconds after the Big Bang. Scientists hope that the explosion will create a swarm of new particles never before detected in experiments. Sifting through the debris of the collisions, cosmologists will look for pointers to which of their ideas about the fundamental properties of the Universe are correct. First on the list is the Higgs boson. Then they will look for signatures of supersymmetry and new dimensions. The latter will manifest themselves, physicists think, with the creation of tiny black holes that live for short amounts of time. Whatever the LHC finds will change physics, bringing the questions surrounding the unification of forces and the mystery of dark matter closer to a solution. But things could also go wrong. What if it doesn't find any trace of the Higgs boson? What if the standard matter particles turn out not to have supersymmetric pairs at the energies that the LHC will be able to get to? Perversely, that has some physicists most excited of all because it would require a complete re-evaluation of the nature of reality.

'Who keeps putting their ready meals in the Hadron Collider?'

As the LHC gears up, some physicists are desperate to make sure that the improbable scenarios are not overlooked. As the physicist Jim al-Khalili wrote: 'If no evidence is found for some crazy idea put forward then it is unlikely that anyone will remember, but if it is vindicated then they'd kick themselves for not having had the courage to propose in advance.'

17 How to age slower than your twin

- Unchanging arenas of action
- It's all relative
- When the speed of light is a special number
- The changing backdrop
- Evidence in the real world
- Energy and mass
- Back to the twins

When Wendy returns from a trip of a lifetime – a trip to explore a nearby star system on a spacecraft that can travel almost to the speed of light – she seems to have fewer wrinkles than her twin brother Daniel, who stayed behind on Earth, and her hair is not as crowded with grey streaks. The date on Wendy's watch, a gift given to her 25 years ago by Daniel when she blasted off, and synchronized with his own, now differs from the date on his watch. To Wendy, only ten years have passed by – Daniel is now 15 years older than his twin.

Unchanging arenas of action

From an everyday perspective, Daniel and Wendy's story is very odd indeed. Whatever tiny disagreement there might have been between their watches before Wendy left Earth, it seems inconceivable that their measurement of the time on her return would disagree by 15 years. From the perspective of the Universe, however, the twins' disagreement is nothing at all out of the ordinary. In fact, the difference exposes how the fabric of space and time really works, a strangeness that is unnoticeable at human scales and only emerges at speeds and energies that are extreme for human experience. We once thought that space and time were the unchanging backdrops to any events in the Universe. Unfortunately not. For anyone who thought it was (and you're not alone, Isaac Newton thought so too), the bad news came in 1905, the 'miraculous year' when a 25-year-old patent clerk in Switzerland overturned our experiential view of nature.

It's all relative

Albert Einstein published five papers in 1905 on a range of subjects, from the electromagnetic force to Brownian motion, which describes the way dust particles move randomly through the air as they are buffeted by molecules of gas. One of these papers introduced the special theory of relativity, a way to relate the motion of different objects as they moved in the Universe.

Isaac Newton had formulated laws of motion more than a century before Einstein, and they worked perfectly well, predicting the motion of everything from cannonballs to planets. They had been proved correct countless times by experiment and there seemed no reason to think they might need improvement. But, while working on the motion of electrically charged particles, Einstein came across a problem with the Newtonian view of the world. His work showed that electromagnetic radiation (such as light) had an upper limit to how fast it could travel and that this speed limit could not be transgressed by anything else. This was problematic for the

Newtonian laws, which had no limits to how fast an object could travel – all you had to do was supply enough energy to keep it accelerating to ever faster speeds. Einstein ended up tearing apart Newton's equations to make them fit the limitations he had discovered. His special theory of relativity smashed the idea that space and time were absolute and fixed features of our Universe. Instead, Einstein said that the values for space and time vary depending on how different observers moved relative to each other.

When the speed of light is a special number

Special relativity is underpinned by two postulates. The first is that the laws of physics are the same wherever you might be. Whether you are in the world's most sophisticated scientific laboratory or the world's most ordinary living room, an experiment (throwing an apple into the air, for example) will have identical results. The second postulate is that the speed of light is a constant, the same value for everyone regardless of their position, or how fast they are moving relative to each other.

This is clearly a strange idea, given how we normally measure speeds. Imagine you're standing on the platform at a railway station as a train pulls in. In the carriage you see a boy passing the time by throwing a ball at the wall opposite him (in the same direction that the train is moving) and catching it as the ball bounces back. If either of you measured the speed of the ball (the 'resting' speed, because the train is at rest) moving away from the boy, you would get the same figure. Now, imagine what happens when the train is moving at a constant speed and, using a telescope, you can still see the boy playing his game. Assuming he threw the ball in exactly the same way as before, someone on the train would measure the speed of the ball as the same as the resting speed. But, from your vantage point at the railway platform, you would see that the ball is moving much faster: at the resting speed plus the speed of the train itself.

The speed of an arrow fired from a moving train towards a stationary target (top) equals the speed of the train plus the speed that the arrow leaves the bow. However, the speed of light leaving a ray gun (bottom) is the same regardless of the speed of the train relative to the target.

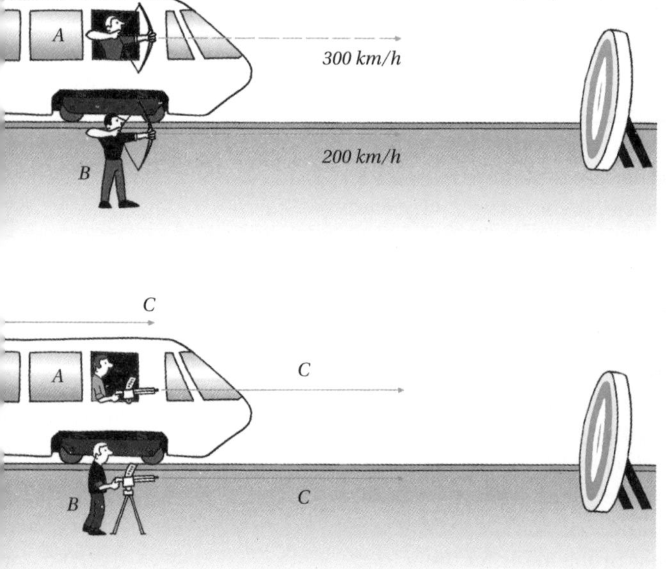

This is not the case with light. Replace the ball with a torch shining in the direction of the train's motion. Whether the train is at rest or moving, whether the person watching it is on the train or on the platform, the speed of the beam of light will be the same – just under 300,000 km per second (186,411 miles per second). With these two postulates, Einstein span out a new world of strangeness.

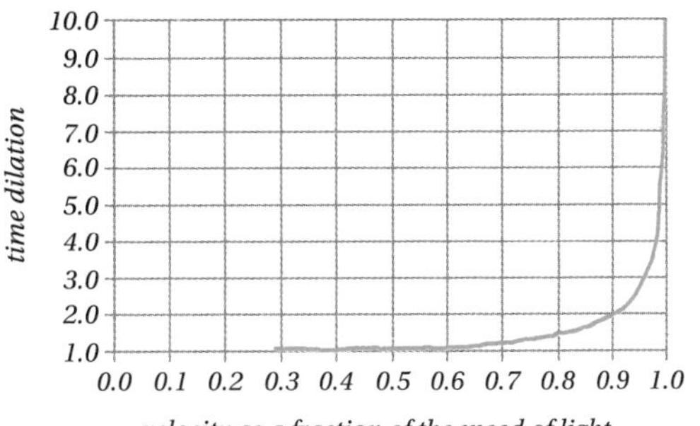

As a person approaches the speed of light relative to another, the moving person's time will appear to the stationary person to slow down (dilate).

The changing backdrop

Speed is measured as distance divided by time. In the example above, where the speed of light is the same regardless of how fast the train is moving, this law of motion seems to break down. But we know that, in special relativity, the laws of physics must remain the same in every frame of reference. So, if our measurement of speed has to stay the same, perhaps the other variables can change? With special relativity, Einstein showed that that the invariability of the speed of light implies a flexibility in the measurement of space and time.

To make the equations of physics carry on working, Einstein showed that the length of any moving object must shrink in the direction of its travel. If the object somehow reaches the speed of light (and we will see later why this is not likely), its length would disappear to zero. Crucially, anyone moving with the object (technically in the object's 'frame of reference') would not notice any change in size; only those observers in a different frame of reference (watching from a train platform, for example, or moving in a different direction) would see the contraction in the object's length. The amount that an object changes in size can be calculated by a mathematical function called the Lorentz transform. Say you have a stick that, when it is in your frame of reference, is 30 cm (12 in) long. Someone steals it from you and puts it on a train that hurtles away at 60 percent of the speed of light. The Lorentz transform will show that, if you could measure the stick, it would only seem to be 24 cm (9.6 in) long.

Just as space is changeable in special relativity, so is time. Until 1905, time measurements were simply a way to put a number on the interval between two events. But Einstein showed that time was, in fact, part and parcel of the coordinates for defining the existence of an event, along with the more familiar coordinates of space. As such, the measurement of time by different observers could differ depending on their relative frames of reference. Just as an observer watching someone else move at a significant proportion of the speed of light relative to them will witness length contraction, they will also see that time slows down for that moving person.

'Einstein never regarded his work on resolving the apparent conflict between classical mechanics and electrodynamics – which led to what we now call the special theory of relativity – as revolutionary in the same sense as his work on the quantum hypothesis.'

John Stachel

Evidence in the real world

The results of special relativity seem bizarre because they're so incongruous with our experience of everyday life. Newton never noticed the need for relativity theory because he never suspected that there was anything special about the speed of light. So he never would have guessed that, as you reach colossal speeds and energies, predicting how things move goes awry. It's true to say that the effects of special relativity are only relevant when speeds get big and that human experience is rarely likely to get there. Even so, scientists have observed its effects in experiments. Particles accelerated to almost the speed of light inside atom smashers experience time dilation – a muon, for example, is known to decay into smaller particles at a very specific rate when it is at rest; but when accelerated, scientists have observed these particles surviving for much longer as time ends up ticking more slowly for them relative to us, in the resting frame of reference.

Energy and mass

The invariability of the speed of light also has some interesting implications for the mass of an object. In short, the faster something moves, the more resistant it becomes to acceleration. Sometimes people explain this by saying that the object's mass increases as it speeds up. But this is not quite right. The object's rest mass (what we measure on weighing scales and is a measure of the matter contained in an object) will stay the same in all situations, but its 'relativistic' mass (which can be used to calculate its resistance to acceleration) increases as it moves faster.

As the object gets closer to the speed of light, the energy needed to make it go faster keeps going up until, at the speed of light, it becomes so resistant to acceleration that no amount of extra energy will make it go any faster. Again, this curious result drops out of special relativity due to the need to maintain the special-ness of the speed of light without, breaking the laws of physics. This part of special relativity also leads to the most famous equation in physics, $E=mc^2$, where E is energy, m is mass and c is the speed of light.

Einstein showed that energy and mass are interchangeable, the latter just being a very concentrated form of the former. Just how concentrated is demonstrated by the value of c^2, a truly enormous number.

Back to the twins

Special relativity neatly explains what happened to Daniel and Wendy. At rest on Earth, they were in the same frame of reference. As soon as Wendy blasts off, they move into different frames that are moving apart relative to one another. As Wendy accelerates closer to the speed of light (and her relativistic mass increases), anyone watching from Earth would see her spaceship get shorter and that time was ticking more slowly on her watch.

Wendy, of course, would be oblivious to any of these changes. For her, life on the spaceship goes on as normal. On her journey, time ticks as it always has done and all the spatial dimensions she sees are the same as they always were. By her measurements, she spends five years on her way to the star, and five years back. Keeping her promise to her brother, she writes in her diary every day and, on her return to Earth, has ten years' worth of entries. The differences between the siblings only emerge when Wendy and Daniel are returned to the same frame of reference. The cumulative effect of the time dilation for Wendy means that she has experienced fewer ticks of the clock during the time she and Daniel have been apart. As such, she returns 'younger' than her brother.

There is one more thing to clear up before Wendy and Daniel's story comes to an end. Special relativity says there is no such thing as absolute motion – frames of reference only move relative to each other. So, if Wendy is moving away from Daniel, isn't Daniel also moving away from Wendy? Looking at it this way, why does time dilate only for Wendy? From her point of view, shouldn't time dilate for her brother as well? Special relativity answers this so-called 'twin paradox' by saying that it is the traveller, Wendy, who defines the actions in this scenario because it is she that leaves Earth and returns. She is the one that leaves Daniel's frame of reference and then returns to it, whereas Daniel has stayed in the same place. She is the one that shifts between different frames of reference and, therefore, she is the one who benefits from staying younger.

Still, it's a curious result. And all because the speed of light has to remain the same at all times, everywhere.

Seen by an observer at rest relative to it, a rocket will seem to get shorter as it reaches a higher proportion of the speed of light.

18 How to get life started

- What is life?
- The chemicals of life
- Where did the ingredients come from?
- Which came first, DNA or protein?
- Did life start in outer space?

Biologists have many ideas about how the simple molecules in the primordial Earth came together and where they might have come from. But it's probably true to say that we know with more certainty about the first few seconds after the Big Bang than we do about the first moments of life. At what point did life actually start? Were the stakes so unlikely that DNA was the only molecule that leapt the hurdles and developed life forms? Or is DNA just the winner among a teeming mass of competitive molecules in the primordial Earth?

What is life?

This question has puzzled even the greats. At the end of his masterwork, *On the Origin of Species*, English naturalist Charles Darwin suggested that the 'creator' had breathed life into a few forms of life, which then evolved into the diversity of organisms we see now. But privately he wondered if a creator was strictly required. In a letter to British botanist Joseph Hooker he speculated that life could have arisen through chemistry alone in a 'warm little pond, with all sorts of ammonia and phosphoric salts, light, heat, electricity etc present.' Is the beginning of life, then, just a matter of the right chemistry?

But what actually defines life? Is it autonomous behaviour? Everything is made of inert matter but life distinguishes itself by moving around, defying whatever fate the dumb laws of physics might have had. Throw a dead pigeon into the air and it will land hard on the ground, governed entirely by the force of gravity and the laws of motion. Throw a living pigeon in the same direction and, though all the same physics applies, the bird will either fly away or land gently. What is the spark of autonomy that gives one collection of molecules (the living bird) such a different behaviour from a virtually identical set of molecules (the dead bird)?

There are other basic characteristics common to all of life: the ability to process energy so that it can function (metabolism); the ability to make copies of itself or have offspring (reproduction); the need for food (nutrition); a complex, organized arrangement of molecules that work together to carry out specific functions and build and control physical characteristics; and the ability to spread or develop (growth). Paul Davies, a physicist at Arizona State University, identifies a few other features. Life also relies, like computers, on the organization of information, which is passed from parent to child through DNA and is forever being modified with random mutations in the genetic code. He adds that all life on Earth is the result of a highly specific deal struck between the 'software' on DNA and the 'hardware' of proteins. Many things in the world have some of these properties, but we would never consider them alive: a fire reproduces, crystals and clouds can grow, and a steam engine needs 'nutrition' in the form of coal to carry out a 'metabolic' process that churns out usable energy. Something that is alive requires all of the traits described above, and for all of those things to be working for the benefit of that organism.

Those examples might be obvious, but examine the definition of life further and Davies warns against looking for a sharp dividing line between living and non-living systems. 'You can't strip away the frills and identify some irreducible core of life, such as a particular molecule. There is no such thing as a living molecule, only a system of molecular processes that, taken collectively, may be considered alive.'

According to Davies, solving the mystery of biogenesis (the creation of life on Earth) is not just yet another problem in the long list of must-do scientific projects. 'Like the origin of the Universe and the origin of consciousness, it represents something altogether deeper, because it tests the very foundations of our science and our world view. A discovery that promises to change the very principles on which our understanding of the physical world is built deserves to be treated as an urgent priority,' he writes in his book, the *Fifth Miracle*.

The composition of the atmosphere has changed throughout the history of Earth. The graph below shows how the proportion of gases has changed over time. The rise in oxygen is a result of the emergence of photosynthesizing plants, which produced the gas as a waste product as they made food from sunlight.

The chemicals of life

There might have been many starts to life, and lots of mechanisms competing with each other to reproduce in the chemical soup of the early Earth. But we know that only one of them, the one based on DNA, won. The confirmation for this is all around us. Trees, vegetables, amoebae, people and whales are all, at the molecular level, based around the same chemistry.

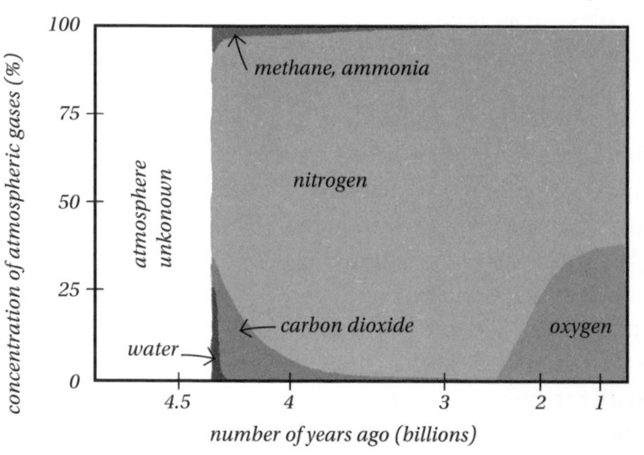

Harold Horowitz, a US biologist and philosopher at George Mason University, has listed the evidence. All life is cellular and all living things are made from 50–90 percent water. Water is the universal solvent for biochemical reactions and is also the source of the key ingredients for photosynthesis, a process in which plants can use sunlight to turn water and carbon dioxide into sugar. The basic molecules of life are the same across all known life-forms (simple sugars, amino acids, DNA, fats) and are virtually all made from largely the same set of 25 elements (mainly carbon, hydrogen, nitrogen, oxygen, phosphorus and sulphur). All replicating cells have a set of genes made from DNA, which is translated into proteins via RNA, a single-stranded nucleic acid. And any physical or metabolic change between generations of an organism (from bacteria to blue whales)is a result of mutations in the genes.

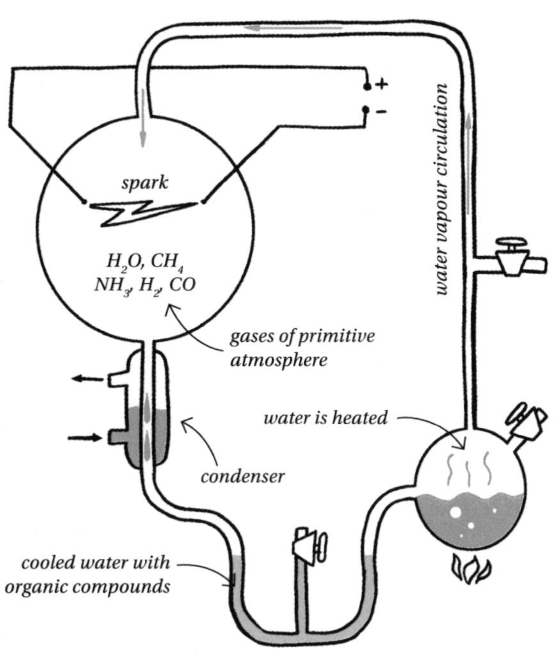

Where did the ingredients come from?

In the 1940s and 1950s, two US scientists at the University of Chicago took inspiration from Darwin's idea of a 'warm little pond' to see if the chemical precursors to life were really present in the soup. Stanley Miller and Harold Urey filled a flask with boiling water, methane, hydrogen and ammonia – at the time these were the chemicals thought to be abundant on Earth around the time life began. They then passed an electric charge through the broth to simulate the lightning strikes that would have occurred naturally in the environment.

Scientists tried to recreate the organic precursors to life by placing gases that were thought to exist in the primitive Earth's atmosphere in a container, and passing electrical sparks through to simulate lightning. The resulting compounds were dissolved and trapped in water, much as they might have ended up in pools of water on an early Earth.

When they opened up the flask, Miller and Urey were astonished to find lots of amino acids (the building blocks of all proteins) and other organic molecules. In variations of their experiment, they produced a rich suite of the other molecules used in life, including sugars, nucleic acids and fats. The Miller–Urey experiments had created, it seemed, some version of the primordial soup.

Nowadays, the results of the experiment are taken with a pinch of salt, given that geologists no longer think that the balance of atmospheric gases on Earth at the time Miller and Urey were simulating were the same as the concentrations in their flask. In addition, later experiments showed that the building blocks of proteins, amino acids, are not actually that difficult to make and even occur naturally in meteorites and in space. But the Miller–Urey work was still interesting, showing that the molecules of life can arise in common chemical reactions.

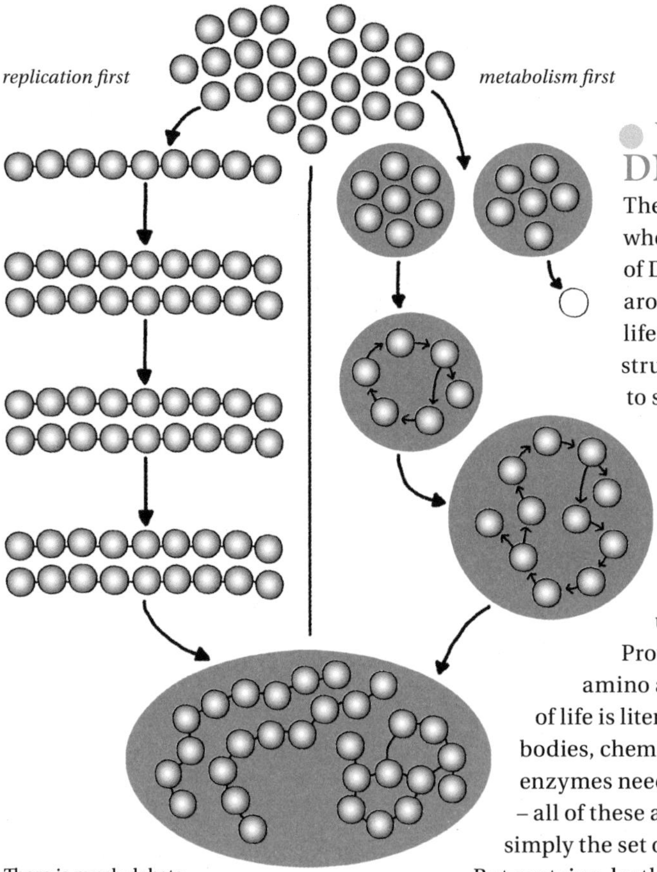

replication first

metabolism first

Which came first, DNA or protein?

There is probably no scientist alive who would not marvel at the brilliance of DNA. This molecule has spread itself around the world, via millions of life forms, evolving all the specialized structures and paraphernalia needed to survive in all of the world's niches. Not bad for a dumb molecule. But one thing DNA cannot do is make anything. By itself it is useless and, if there were no proteins around, DNA would just have made endless copies of itself until the ingredients ran out. Proteins, which are long chains of amino acids, are the stuff that all the form of life is literally made from. Eyes, petals, bodies, chemical messages between cells, the enzymes needed to speed up chemical reactions – all of these are made from proteins. DNA is simply the set of instructions to make proteins. But proteins, by themselves, are useless at a vital part of creating new organisms – reproduction. They cannot spontaneously copy themselves. For that, they need DNA. The question that has long-vexed biologists is, which came first?

There is much debate around which of the fundamental components of life came first, whether it is the replicating element (such as DNA or RNA, on the left of the diagram), which copies life from one generation to another; or the metabolic element (on the right), which keeps an organism functioning during its life.

The most promising idea is that DNA evolved from its simpler cousin, RNA. This molecule, abundant in living cells today and used to read the genetic code and assemble amino acids into proteins, can also replicate itself. Perhaps RNA assembled the first proteins from the primordial soup, getting ever more complex itself in the process and eventually developing, via mutations, into DNA. It's a thought, anyway, but still unproven.

Did life start in outer space?

Even though Miller and Urey showed that the chemical precursors to life could quite easily have come about in a warm pool, what would eventually cause the mass of molecules to come together into a living being? Would an ocean of watch-parts ever spontaneously create a watch, for example, even

'The climate and the chemical properties of the Earth now and throughout its history seem always to have been optimal for life. For this to have happened by chance is as unlikely as to survive unscathed a blindfold drive through rush hour traffic.'

James Lovelock

if left alone for millions of years? Perhaps, instead, life started on Earth because it came, ready-formed and with instructions on how to use the ingredients in the primordial pools, from space. It sounds, at first blush, a bit ridiculous. But this idea has been at the back of scientists' minds for decades. Swedish scientist Svante Arrhenius proposed 'panspermia' in 1908, that is the seeding of the ingredients for life from meteorites and comets that have hit Earth. These interplanetary travellers could carry molecules of even micro-organisms from one world to another. Even earlier, in 1821, Frenchman Sales-Guyon de Montlivault had proposed that life started with seeds from the Moon.

So far we have assumed that all the necessary requirements for life were on Earth before that life emerged. Is panspermia a possibility? The available evidence does not rule out the idea entirely. Bacteria that can survive the deep freeze, vacuum and harsh radiation of outer space already exist here on Earth. Small colonies of *Bacillus subtilis* can survive being electrocuted, exposed to vacuum and cosmic rays. In one experiment, most of a sample of tobacco mosaic virus remained infectious after being cooled to –196°C (–321°F) and bombarded with protons for a day. The *Micrococcus radiophilus* bacterium has incredible resistance to radiation, having developed impressive mechanisms to repair its DNA in the event of damage by X-rays.

If life developed on one planet, goes the panspermia hypothesis, then what's to stop it travelling to another planet via interstellar dust clouds or, more likely, meteorites. Chemical analyses of meteorites on Earth has indeed shown that they sometimes contain complex organic molecules and sometimes even what scientists in the 1960s once called 'organized elements' that may have been biological in origin. Alas, it remained unproven. And no-one has yet found the definitive evidence for a microbe (alive or fossilized) in any object from outer space, which would answer whether or not panspermia is correct once and for all. Mind you, a discovery of microbes from space would probably make headlines for another reason entirely.

19 How to predict the unpredictable

- The emergence of chaos
- Chaos, chaos everywhere
- The limits of chaos
- Strange attractors
- How chaos theory becomes useful

Unpredictability keeps coming back to haunt scientists. In the second half of the 20th century, an idea came about that sounded like it could perhaps never be tamed: chaos theory. This tells us that, though everything around us is built on the rules of physics, those simple rules have managed to create a world of such dense and ever-changing complexity that it is almost impossible to predict. If we do try to model it, the tiniest error in that model could lead to a meaningless result. Is this finally the end of predictability for science?

The emergence of chaos

If there has been a unifying theme in the natural philosophy that emerged from the enlightenment, it probably had something to do with the idea that revealing the structure and laws of nature would give us some control of our fate. 'If we know the pulleys by which things act,' went the overriding sentiment among scientists, 'then we can, given the right starting conditions, predict everything about its future. Nothing will be a mystery for much longer.' At the start of the 21st century, we live in a world that has made good on some of that promise – the laws of physics discovered by Isaac Newton and James Clerk Maxwell still work and they still enable us to do remarkable things. But scientists are no longer content in thinking that we can predict everything like clockwork. In the past century, scientists have found unpredictability in natural things that none of their great predecessors could have suspected.

'As a boy I was always interested in doing things with numbers, and was also fascinated by changes in the weather.'

EDWARD LORENZ

Enter chaos theory. The startling effects of chaos are usually described by something called the 'butterfly effect'. This familiar tale has it that the flap of butterfly wings over Texas could result in a storm over Japan. You can pick whichever places you like for the locations of the butterfly and the storm, but the message is clear. A tiny event (the butterfly moving a miniscule amount of air) can build up through lots of knock-on effects into something truly startling thousands of kilometres away. It sounds a little fanciful and, of course, millions of butterflies flap their wings thousands of times a day but there aren't billions of storms every day. Which means that, in most cases, butterfly wings do not lead to anything catastrophic at all.

The point of the metaphor is that the progression of weather (and, indeed, any chaotic system) is very sensitive to its starting conditions. A weather system might sound like it should be simple to characterize – after all, it is just atoms and molecules at different temperatures floating and interacting in different fluids. The physics of every process that each molecule goes through is well understood. In Isaac Newton's clockwork world, that information should be enough to predict the path of every molecule.

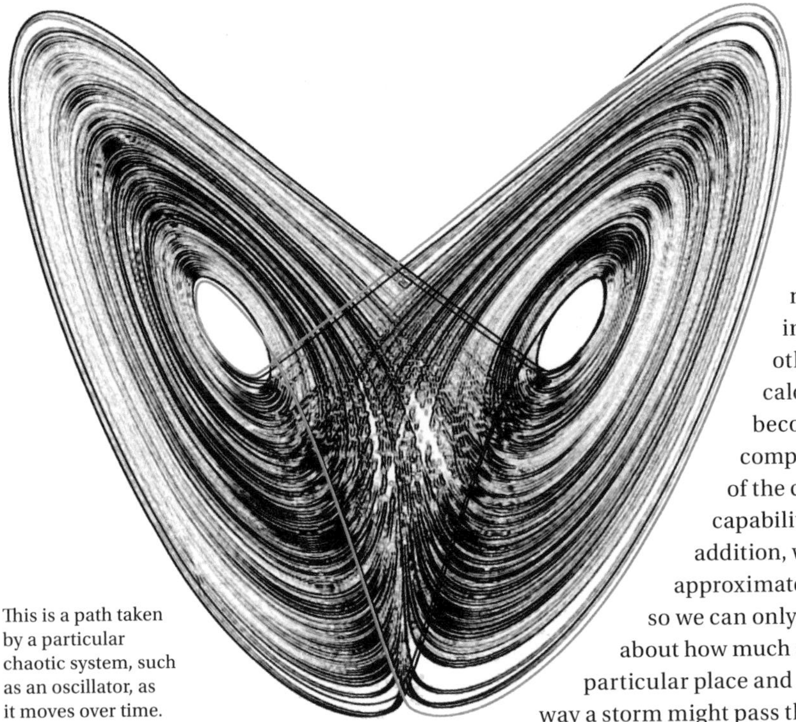

But there are so many molecules, each interacting with so many others around it, that the calculations very soon become fiendishly complicated and spiral out of the control of any human capability to understand. In addition, we can only know approximate starting conditions, so we can only have approximate ideas about how much sunshine will fall in a particular place and only approximate the way a storm might pass through a city. Because there are so many variables and particles involved and because the starting data is so approximate, any errors become compounded. The results of the thousands of steps of calculations that a weather simulation needs to do can therefore become meaningless. (In fact, meteorologists have quipped that any computer model sophisticated enough to truly predict the weather would have to be as complex as the weather itself – in short, the only thing that can simulate weather is, well, the weather.)

This is a path taken by a particular chaotic system, such as an oscillator, as it moves over time. The oscillator traces this path out in three dimensions and, though it never takes the same path twice, it draws out a recognizable pattern.

Studying the weather is how chaos theory first came into existence. US mathematician Edward Lorenz had been working as a weather forecaster for the United States Army Air Corps during the Second World War. In the 1960s, he tried to develop a computer model for weather and noticed that even the slightest change in starting conditions for the dozen or so variables he used resulted in huge differences in the weather that the model would go on to predict. Lorenz is acknowledged as the father of chaos theory, and it was he who came up with the idea of the butterfly effect.

Chaos, chaos everywhere

Once you know it exists, chaotic behaviour becomes apparent in all sorts of places. Consider the Solar System, which Newton observed and which inspired him to formulate many of his ideas about the clockwork universe.

Scientists are not lying when they say they can confidently work out the orbit of Earth around the Sun. All you need is some knowledge of the gravitational forces acting between the two bodies and you have an answer. But, as any schoolchild knows, the Solar System contains more than just the Sun and Earth. There are seven more planets, scores of moons and millions of other bits of asteroid and rock. Each bit has a gravitational impact on every other bit – all of them in exactly the way Newton described in the 17th century. Good luck trying to solve the mathematical equations that are required to describe every bit of this multi-part system. Henri Poincaré long ago proved that, while there are solutions to the equations that describe the interaction between two bodies, anything more than that is virtually impossible to solve. That system becomes chaotic.

The limits of chaos

All this talk of unpredictability can get depressing. If we can't use the laws of physics to do the predicting we once hoped for, what good are they? What use are the laws if they are impractical to use in any real-world situation?

Fortunately, the study of chaos has revealed some handles that we can grab onto as we swirl around. Although the orbits of the planets in the solar system are chaotic, there does seem to be a limit to the range of paths they take in the maelstrom. It may be impossible to know exactly which line a planet or asteroid will follow around the sun, but we can predict some properties of what might happen. For example, planetary orbits are bounded into a range, as if they were attracted to a particular path that they never quite achieve. Round and round they go, each orbit subtly different from the one before it; watch the planet for long enough and you'll notice a pattern.

A double pendulum is a simple-looking system that displays chaotic behaviour as it moves back and forth.

Strange attractors

If the idea of chaos arising in simple systems is not strange enough, the properties of those chaotic systems might give you more cause to raise an eyebrow. One of the simplest examples of a chaotic system is the double pendulum. This is a system a bit like your leg, with two joints around which two rigid rods can swing freely. Unlike your leg, of course, the middle joint can swing both ways. If you get a double pendulum swinging, it might behave fairly predictably for the first few swings, perhaps you would notice nothing different from a normal pendulum, a device so predictable that it is used to teach generations of children about some of the basic properties of oscillations and waves.

But sooner rather than later things will go haywire. The pendulum will move wildly from one side to the other, with the length of each swing highly irregular and every movement looking almost random. The bottom of the lower pendulum stops just swinging from side to side and starts looping around gymnastically. Watch one of these in real life and you will soon begin to wonder if there isn't something or someone controlling the pendulum somehow.

Fractals are complex geometric shapes that have infinite detail. Zoom in on a section as much as you like and you will see ever more detail that looks exactly like the original geometric shape. Fractals are related to chaos because they are based on simple iterative equations, where each step can be easily understood but the collective result of billions of steps becomes very complicated.

If you were to watch the bottom part of the pendulum only, however, and trace out the path it follows in space, you would notice something interesting. Though it looks random and though the pendulum never repeats a loop twice, it is as if the bottom of the pendulum is doing its best to draw out a specific pattern. This property of chaotic systems, dubbed the 'strange attractor', allows anyone studying them to at least get a start on how a system might behave.

How chaos theory becomes useful

The strange attractor is the start of how chaotic systems can be tamed for studying the complex systems in nature – whether in biology, the environment or the stock market. Chaos theory has been used to study ecological systems, for example. A population of animals might grow quickly when food is readily available, then crash as the food disappears after a drought and cannot sustain so many creatures. Predicting populations had always been difficult, however, because the simple rules sometimes have wild spikes or drops in population that could seemingly not be explained. Chaos theory helped mathematical biologists to work out that these wild changes were not a problem with the models but could, in fact, just be part of the natural system.

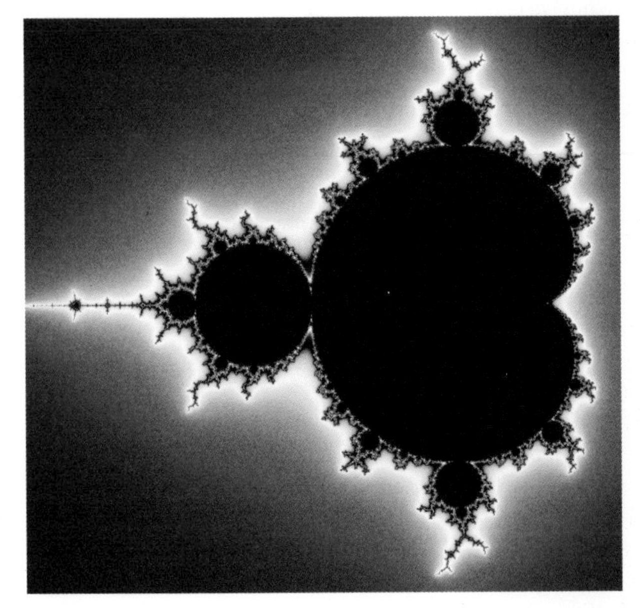

In fact, chaos theory is leading to other insights about the ever-changing competitions between different species. Some scientists believe that two species coupled in some way in an environment will evolve together so that they exist just on the transition point between order and chaos. Imagine a pond that contains frogs and flies. If the flies are a source of food for the frogs, then natural selection will mean that the flies with the best strategies for evading the sticky-tongued amphibians are most likely to survive. In response, the frogs with the best strategies for catching the ever-more-clever flies will do best at surviving too. In effect, the two species set up an arms race – in chaos terms, they are 'coupled'. If the frogs and flies are not very strongly coupled in the pond, then the changing strategies of one will not effect the other so much – the pond is in an ordered, static state. If, on the other hand they are very strongly coupled, then even a slight change in one species' strategy will result in wholesale and random changes in populations – in effect the pond will be in a chaotic state with boom and bust all the time. This is no good for either species.

Something interesting happens, however, if the frogs and flies have an interaction that is not too strong and not too weak, which is much more likely in a real-world situation. Then they can reach a critical point that is just on the edge of chaos – allowing the species living there to change strategies subtly without it resulting in huge spikes or crashes in population every time. Some chaos theorists believe that in coupled environments, such as our pond with frogs and flies, a species reaches optimum fitness when it evolves to sit on the border between order and chaos.

The mathematics of chaos theory can, some scientists say, also be used to understand the complexity we see around us. In the 1970s, even before chaos became a popular field of study, scientists Ilya Prigogine, Herman Haaken and Herbert Simon were looking for general principles for systems where many parts interacted and communicated to give complex results. The things they wanted to explain included what they called adaptive systems – cells, whole organisms and even financial markets. These are systems capable of forming strategies in response to environmental cues – an oak tree might orient its leaves to the sun, for example, or the market for pork bellies could respond to an excess of product by lowering prices. All of these systems work on relatively simple rules at the tiniest scales, and manage to coordinate without any central intelligence. Chaos theory (perhaps strange attractors or phase transitions between order and chaos) may help to explain what is going on.

It might have a name that makes scientists sound like they've given up. In truth, though, chaos theory has shown itself to be one of the most useful ideas when it comes to predicting unpredictable things.

20 How to fight for survival

- Natural selection
- The evolution of natural selection
- Strategies for success
- Survival in the 21st century

Nature is brutal. Its daily ritual is disease, suffering and death. Species do whatever they can to survive, taking advantage of any random physical changes that might have accrued over generations of their kind, changes that perhaps make them better-adapted to their environment than their cousins or neighbours. That process of change, evolution by natural selection, is the grammar of the biological world, the rules that become apparent when you watch life assemble itself, change or get destroyed over millions of years. It might seem cold, nasty or even wasteful. But the logic is undeniable.

Natural selection

Charles Darwin's theory of evolution by natural selection is one of those ideas in science that, like all the greatest discoveries, does much more than explain the Universe – it changes how we see ourselves in that Universe. That species change over time is undisputable – the fossil record shows us countless strange creatures that no longer exist and yet more that we can trace back as having something to do with modern elephants, tigers or us. Evolution is fact. How species change was more of an open question before Darwin came forward with his idea. Natural selection is so simple that you can summarize it in just three sentences.

Each generation of life passes hereditary information to its offspring. Occasionally that information is randomly mis-copied, a change that can subtly alter the ability of a species to survive in an environment. Natural selection is the result of the interactions between those changed species in a particular environment as they fight for survival. Over millions of years, natural selection has led to all of the diversity and complexity we can see in nature. If you want to reduce Darwin's idea still further, you could say it is simply 'descent with modification'.

The evolution of natural selection

Darwin was a model scientist. Natural selection is an idea that might seem so obvious that it could have come to him in a flash, rather like how Isaac Newton was said to have been inspired to formulate gravity when he saw an apple fall from a tree. From that one act, Newton went on to describe the way the Universe was put together, and to predict the motions of planets and stars. Not so with natural selection. Darwin was a patient collector of data and had already started his natural history work before joining the voyage of the *Beagle* to the New World in 1831. The trip took him to the continents of South America and Australia, where the young Darwin continued his study

of as many species of animal and plant he could lay his hands on: drawing them, spotting patterns, noting differences. His collection of data grew but it took him more than two decades of painstaking study and deliberation (via several years and volumes of work on barnacles) before he was ready to reveal his thesis. *On the Origin of Species* was published in 1859, though the ideas of natural selection had been presented to the world a year earlier.

Darwin's 'finches' are a group of 15 birds of the tanager family found on different islands of the Galápagos. The biggest difference between them is the size and shape of their beaks, each adapted to the different food sources in their habitats.

Naturalists knew that species existed, of course, and they knew they competed for resources. But, until Darwin, they did not have a way of fitting everything together, of working out how or why things were related, or getting a handle on why things kept changing over time. Darwin started *Origin* with the question of where species came from, since his first task was to deal with the prevailing idea that creatures came ready-made and were largely unchangeable. 'Who can explain why one species ranges widely and is very numerous, and why another allied species has a narrow range and is rare? Yet these relations are of the highest importance, for they determine the present welfare, and, as I believe, the future success and modification of every inhabitant of this world,' he wrote.

On the Origin of Species was an explosive hit for its time, selling more than 100,000 copies in various forms by the turn of the 20th century. It gave a structure to biology that had been sorely lacking, allowing its students to start asking questions about why some species survive and others don't.

It is worth saying that Darwin never meant natural selection to mean 'survival of the fittest', the common shorthand for his idea. His idea was about the competition between and survival of traits and individuals within a species, rather than between species. Minute changes between successive generations mean that some members of a species are better than others at surviving. Perhaps they can eat more of the berries around them or they have stronger claws. They are more likely to live long and end up having offspring. We can also deal with another misconception: that natural selection implies humans are descended from monkeys. Darwin never said that and the error is a profound misunderstanding of natural selection, akin to saying that you are the child of your cousins. In actual fact, Darwin said that monkeys and humans must have had a common ancestor at some point in history (millions of years ago) and that we both evolved from that ancestor along different paths.

Strategies for success

The study of natural selection took off in the 20th century when scientists began to investigate the different strategies that species use to maximize their fitness in an environment. All of these strategies are, to a large extent, passed on from generation to generation, with the best competing to gain supremacy within a species. A creature might live in a territory alone or in groups that all share the task of raising young. It might display self-interest when hunting for food or it might altruistically share what it had found with all members of its pack. All these behaviours had to be explainable via natural selection and biologists began assembling the bits of biology into the overall framework of Darwin's theory.

The advent of mathematical techniques in the 1970s was of particular help, allowing biologists to experiment with virtual species inside computer simulations, each of which could be programmed to employ different strategies for survival. A simulation might contain 'organisms' that have to share resources in a virtual environment. Each organism has defensive strategies (encoded in 'genetics' that they can pass on, in a modified way, to their offspring) and wanders around eating and fighting. If it survives long enough, an organism can produce offspring that can develop even more elaborate defensive techniques as the successive generations go by.

The simplest strategy that these virtual organisms can be programmed with is in a simulation called 'chicken' – after the children's game of the same name. If two animals come across the same piece of food, they can fight for the food or let their opponent have it. The advantage of the first strategy (called 'hawk') is that there's a chance the animal will eat, the disadvantage is that they might get injured or die. With the second strategy (known as 'dove'), they are certain not to get hurt, but they are also certain to go hungry. A 'hawk' in the midst of 'doves' will win every fight and produce more offspring that will continue to

win fights. Eventually, hawks will outnumber doves in that population. However, because hawks fight, they will necessarily start killing each other, so their populations will stabilize or even fall over time, allowing space for the doves to increase their numbers as the hawks wipe themselves out. And so the cycle repeats. By looking at how the populations of an organism change over time in a simulation, biologists can compare their computer models to data from the real world and try to work out which strategies the species out there are using.

Another game is called 'tit for tat'. This takes the basic chicken game on a few levels by adding two common behaviours seen in animals – co-operation and learning – and is a modified form of the classic prisoners' dilemma. In that problem, two criminals are interrogated. They are separately offered a deal: if one confesses and the other doesn't, the confessor will be set free and the other jailed. If neither confesses, they are both set free. But if both confess, both go to jail. Clearly, the prisoners would do well to co-operate and stay quiet. But, without a means of communication, how does the silent one know that the other is also staying quiet?

In nature, this dilemma is played out in all sorts of situations. Two organisms may benefit from working together but one may get more out of the bargain if they do less of the work. Tit for tat takes into account any laziness since it allows organisms to change their actions depending on the behaviour of the partner. An animal might co-operate the first time it meets the other but then only co-operates a second time if it got a good result the first time. This reduces the number of fights that an animal will get into, while giving it the possibility of help on some occasions.

Survival in the 21st century

Darwin knew nothing about genes, but it is a credit to his incredible theory that every subsequent generation of scientists has found evidence to confirm his ideas. Natural selection requires hereditary information to be passed between generations (sometimes in modified ways), so there had to be some way of reflecting strategy and suitability to an environment in an animal's biology. In the middle of the 20th century, that question was answered by the work of, among others, US scientist James Watson and British scientists Francis Crick and Rosalind Franklin, which showed that DNA was the primary carrier of the hereditary information.

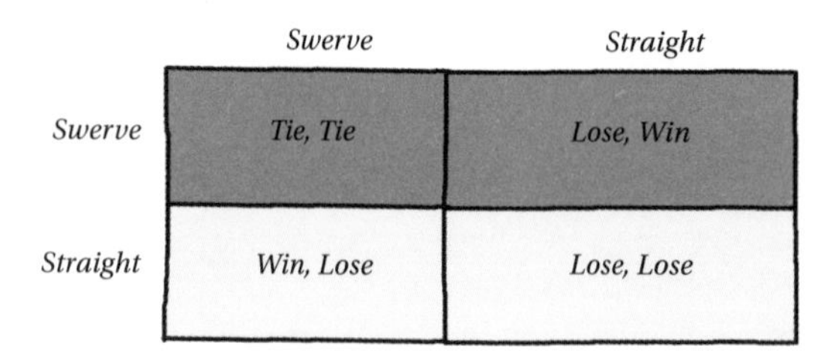

	Swerve	*Straight*
Swerve	*Tie, Tie*	*Lose, Win*
Straight	*Win, Lose*	*Lose, Lose*

This 'payoff matrix' shows the different possible outcomes of the chicken game, depending on the actions of each player. Two players are running towards each other (or aiming to get a piece of food) and the winner is the one who carries on moving straight and doesn't swerve out of the way. Each player would prefer to win over tying, tie over losing and to lose over crashing into each other.

'The fact that life evolved out of nearly nothing, some 10 billion years after the Universe evolved out of literally nothing, is a fact so staggering that I would be mad to attempt words to do it justice.'

Richard Dawkins

Mutating nucleotides (the letters in your genetic sequence, which influence your physical features, general abilities and behaviours) can have one of several results. Sometimes nothing at all happens – what's one change in billions of letters, after all? Sometimes mutations can lead to problems – a single defect can lead to cystic fibrosis, for example, or other genetic diseases where an individual ends up dying before they can pass on their genes. And, other times, a mutation makes an individual more adapted to take advantage of its environment. A mutation can infer immunity against a pathogen (a virus or a bacterium, say), making it live longer and more likely to reproduce. Closer to home, mutations in human DNA have led to some remarkable advantages for our species. The HAR1 sequence of gene is involved in giving us our unusually large brains – there are two changes between the HAR1 genes of chickens and chimpanzees but 18 changes between chimpanzees and us. Similarly the FOXP2 gene, involved in producing communication in many animals, seems to have undergone a developmental acceleration in humans.

Laurence D Hurst, professor of evolutionary genetics says that the evolutionary changes in an organism's DNA are like a race between two drunk people:

> *At each step, each drunk is as likely to walk forwards as backwards. Place our drunks on the start line of a 100 metre track and add one extra rule. A drunk who steps back over the start line is out of the race. The drunks in the analogy are random changes to DNA and their backwards and forwards steps represent changes in the frequency of those mutations in a population with each generation. Because they begin close to the start line, most drunks will be quickly eliminated. But some, by chance, will make it to the finish, not because they were good runners but because they happened to go forwards more often than backwards. This is the neutralist view of evolution – lucky genes, not selfish genes.*

Tiny changes at the molecular level shoot off in every possible direction, with natural selection paring down the most suitable changes into viable organisms. Only when an organism can take advantage of the luck of its mutations, the changes in the abilities that its biology gives it by chance, only then will it survive.

21 How to boil a planet

- Moving energy around the world
- The greenhouse effect
- How Earth is warming
- Could natural variability cause climate change?
- Consequences of the warming world
- Can the world curb its emissions?

Earth's climate is the long-term result of complex interactions between the atmosphere, oceans, living organisms, ice sheets, soils and rocks. The result of those interactions is familiar to us as weather – everything from air temperature to storms, rain and snow to the height of the sea around coasts. Because of the industrialization of the world over the past 150 years, we have been irrevocably altering the balance of energy that flows into, around and out of our world. We humans have been slowly cooking our planet.

Moving energy around the world

The key factor in determining climate on Earth is the balance between how much energy arrives from the Sun versus how much leaves our planet. A quarter of the energy is absorbed by the clouds, a quarter is absorbed by the atmosphere and the rest travels to the surface. The oceans store vast amounts of heat and distribute it around the planet via major currents, which transfer energy from warm equatorial regions to the colder polar latitudes.

Earth reflects and radiates away most of the energy it gets from the Sun. Ice sheets in Antarctica, the Arctic, Greenland and on most of the world's highest mountains reflect away sunlight directly. The darker parts of Earth, such as oceans or forests, absorb the Sun's energy and the latter radiate it away as infrared radiation, or heat. Some of this heat is kept in thanks to the natural greenhouse effect of the atmosphere, which keeps Earth's surface temperature at an average 15°C (59°F), as opposed to the –18°C (–0.4°F) it would be if there was no atmosphere at all. Overall, this natural energy balance keeps climate relatively stable on Earth. But humans have inadvertently been tinkering with it since the middle of the 19th century. Burning fossil fuels and emitting copious greenhouse gases, such as carbon dioxide and methane, into Earth's atmosphere, has upset the planet's energy balance. And that means temperatures have been on the rise.

The greenhouse effect

Earth's atmosphere is made of mostly nitrogen (78 percent) and oxygen (21 percent) plus a mixture of trace gases, including carbon dioxide and methane. Despite forming less than 1 percent of the atmosphere, these two gases, plus a few others in even smaller amounts, play a disproportionate role. The 'greenhouse effect' was first described by the Swedish chemist Svante Arrhenius at the beginning of the 20th century. So-called greenhouse gases, because of their molecular structure, absorb some wavelengths of the high-energy electromagnetic radiation from the Sun but allow others

The Sun is the sole external provider for energy for Earth. Some of its energy is deflected back into space, a quarter is absorbed by the atmosphere, a quarter is absorbed by the clouds, and the rest travels to the surface. The greenhouse gases are spread throughout the atmosphere rather than concentrated into a shell around the world.

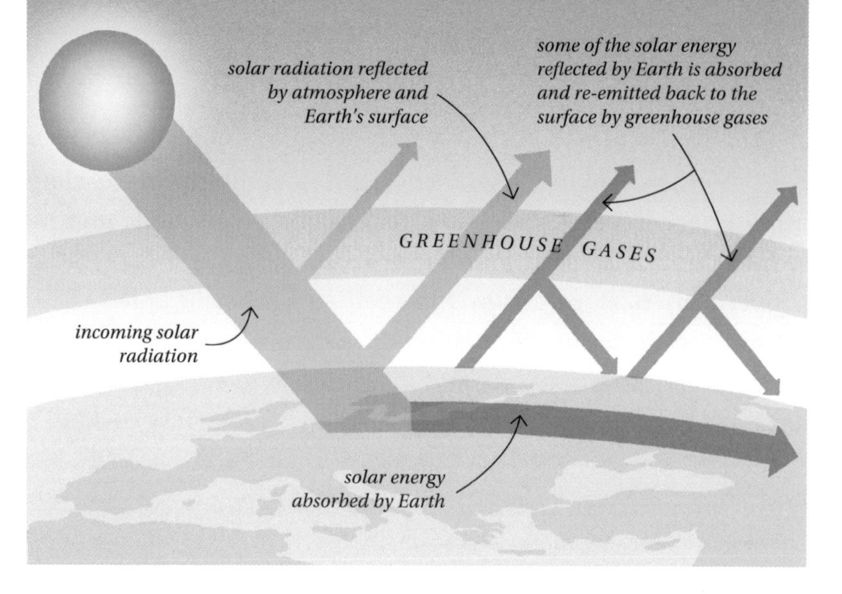

(mainly visible light) to pass right through. These gases can also trap some of the heat that is radiated towards space from the surface of Earth, making the surface warmer than it would be if there was no atmosphere.

More greenhouse gases in the atmosphere will trap more heat near the surface. According to the Intergovernmental Panel on Climate Change (IPCC), global concentrations of greenhouse gases increased by 70 percent between 1970 and 2004. Additionally, 'Global atmospheric concentrations of carbon dioxide, methane and nitrous oxide have increased markedly as a result of human activities since 1750 and now far exceed pre-industrial values determined from ice cores spanning many thousands of years.'

How Earth is warming

We know climate change is happening because it is a scientific discipline that is rich with data. Satellites, airborne and ground-based measuring stations and expeditions to remote parts of the world record everything from rainfall to air and surface temperature, ice thickness to humidity. According to NASA, the past decade has been the warmest on record, and 2009 the second warmest year. Temperatures have risen by 0.2°C per decade over the past 30 years and average global temperatures have increased by 0.8°C since 1880. The data paints a clear picture: the average surface temperature of Earth is rising.

In its 2007 report, IPCC said that average northern hemisphere temperatures in the second half of the 20th century were very likely to be higher than in any other 50-year period in the last 500 years and perhaps at least the past 1,300 years. Spring was getting earlier every year and animals and plants gradually shifting northwards as parts of the world became too warm for them. In the oceans, algae, plankton and fish were all moving away from warming water.

'Future generations may well have occasion to ask themselves, "What were our parents thinking? Why didn't they wake up when they had a chance?" We have to hear that question from them, now.'

Al Gore

A lot of the latest evidence has come from new measurements in Earth's oceans. More than 80 percent of the heat trapped in the climate system as a result of increased greenhouse gases is exported into the oceans, and scientists have recorded this happening. Another impact is a change in water salinity: as the atmosphere warms up, more water evaporates from the surface of the ocean, leaving the remaining water more salty. This has been most noticeable in the sub-tropical Atlantic. Then there is the parlous state of Arctic sea ice, which is retreating at an ever-increasing rate – UK climate scientist Peter Stott of the Met Office calculated that an area the size of Madagascar has disappeared every decade. Rainfall is also on the rise in the higher latitudes of the northern hemisphere and large swathes of the southern hemisphere while, in the tropics and sub-tropics, it has fallen.

Could natural variability cause climate change?

Greenhouse gases are not the only possible natural causes for climate to change. The Sun's energy output follows a cycle, rising and falling over 11 years, and some think this could account for the changing temperatures of the past decades. It sounds sensible, but has been widely discredited, as Stott's review proved: 'There hasn't been an increase in solar output for the last 50 years and [it] would not have caused the cooling of the higher atmosphere and the warming of the lower atmosphere that we have seen,' he said. If climate change was entirely due to solar activity, Earth's atmosphere would have warmed more evenly – affecting both the troposphere and stratosphere – and temperatures should have increased more quickly earlier rather than later in the 20th century, which is the reverse of what was actually measured.

Consequences of the warming world

Rising temperatures on Earth will have some devastating impacts on all life. If all the dire predictions about climate change come true, what can we expect to experience as the planet starts to boil? In his book, *Six Degrees*, British environmentalist and author Mark Lynas put forward a series of scenarios as temperatures go up, impacts that can be corroborated by the

1950

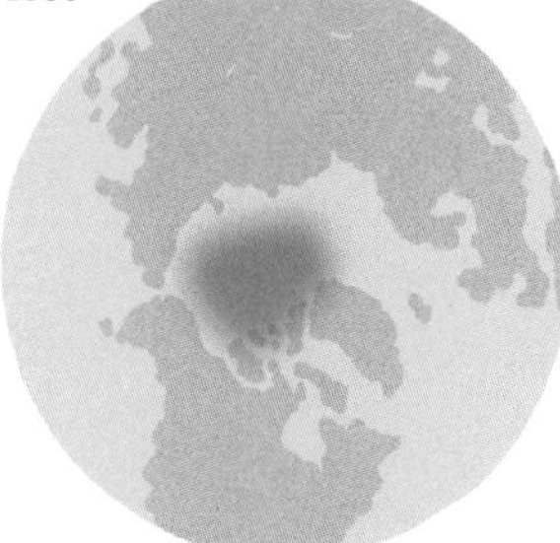

2000

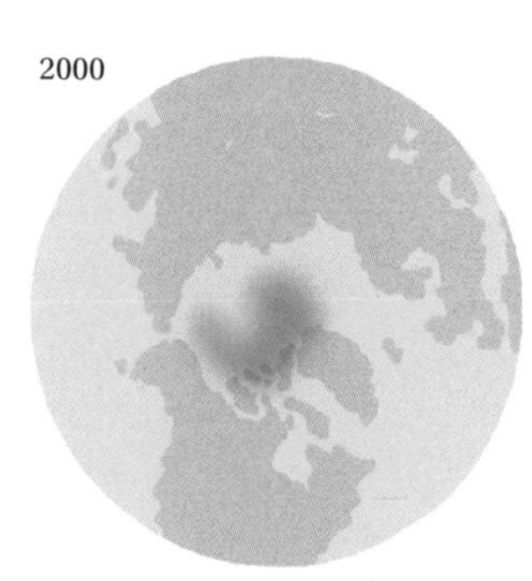

2050

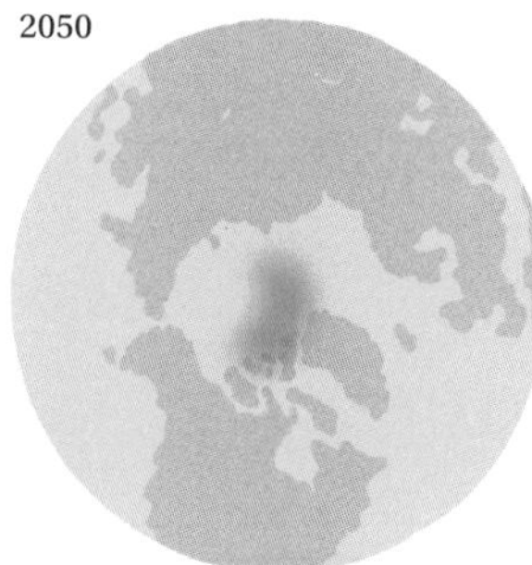

The extent of permanent sea ice over the Arctic Ocean has retreated since 1950, and is predicted to continue to retreat as temperatures rise.

scientific evidence assessed by the IPCC and the Met Office. With a 1°C (1.8°F) temperature rise, Arctic sea ice will disappear for good in the summer. Heatwaves and forest fires will become more common in the sub-tropics with the Mediterranean region, southern Africa, Australia and the south-west US worst hit. Most of the world's corals, including the Great Barrier Reef, will die. Glaciers providing fresh water for crops that feed 50 million people begin to melt and 300,000 people are affected every year by climate-related disease, such as malaria and diarrhoea.

With a 2°C (3.6°F) rise, heatwaves of the kind seen in Europe during 2003, which killed tens of thousands of people, will be common. The Amazon will turn into desert and grasslands, while increasing CO_2 in the atmosphere make the oceans too acidic for remaining coral reefs and thousands of other marine life-forms. More than 60 million people, mainly in Africa, be exposed to higher rates of malaria and agricultural yields around the world will drop, leaving half a billion people at greater risk of starvation. The West Antarctic and Greenland ice sheets melt and the world's sea level will rise by 7 m (23 ft) over the next few hundred years. Glaciers recede, reducing the fresh water supply for major cities including Los Angeles. Coastal flooding affects more than 10 million extra people. A third of the world's species become extinct as the warm climate changes their habitats too quickly for them to adapt.

With a 3°C (5.4°F) rise from today's levels, global warming may run out of control and efforts to mitigate it may be in vain. Millions of square kilometres of Amazon rainforest could burn down, releasing carbon from the wood, leaves and soil and making the warming even worse, perhaps by another 1.5°C (2.7°F). In southern Africa, Australia and the western US, deserts take over. Billions of people are forced to move from their agricultural lands in search of food and water.

At a 4°C (7°F) rise, the Arctic permafrost enters a danger zone. The methane and carbon dioxide currently locked in the soils will be released into the atmosphere. At the Arctic itself, the ice cover will disappear permanently, meaning extinction for polar bears and other native species that rely on the presence of ice. Further melting of Antarctic ice sheets would mean a further 5 m (16 ft) rise in sea level, submerging many island nations. Italy, Spain, Greece and Turkey become deserts and mid-Europe reaches desert temperatures of almost 50°C (120°F) in summer. Southern England's summer climate could resemble that currently seen in southern Morocco.

With a 5°C (9°F) rise, global average temperatures will be hotter than for 50 million years. The Arctic region will see temperatures rise way above average – up to 20°C (68°F) – meaning the Arctic is ice-free all year round. Most of the tropics, sub-tropics and even lower mid-latitudes are too hot to be inhabitable. Sea level rise is so severe that coastal cities across the world are abandoned.

Can the world curb its emissions?

In November 2009, a team of scientists at the University of East Anglia (UEA) predicted that the world is on course for 6°C (11°F) of global warming. More worrying, Earth's natural ability to absorb the gas seems to be declining. With the British Antarctic Survey, Corinne Le Quéré of UEA studied 50 years of data on carbon emissions, including estimates of human emissions and other sources such as volcanoes. The team also estimated how much CO_2 was being absorbed naturally by forests, oceans and soil. Those natural sinks, they found, are becoming less efficient, absorbing 55 percent of carbon, compared with 60 percent half a century ago. This drop is equivalent to 405 million tonnes of carbon. Le Quéré also showed that CO_2 emissions from burning fossil fuels increased at an average of 3.4 percent a year between 2000 and 2008, compared with 1 percent a year in the 1990s. The vast majority of the recent increase has come from China and India, though a quarter of their emissions are a direct result of trade with the West. In recent years, the global use of coal has also surpassed oil.

Limiting the average temperature rise to 2°C (3.6°F), which scientists say could avoid the worst effects of climate change, would require CO_2 emissions to peak between 2015 and 2020 and that the global per capita emissions be decreased to 1 tonne of CO_2 by 2050. Currently the average US citizen emits 19.9 tonnes per year and UK citizens emit 9.3 tonnes.

Publishing her team's work just before the 2009 UN summit in Copenhagen, Le Quéré called for governments to adopt targets to reduce greenhouse gas emissions to limit the global temperature rise to 2°C, or face the real threat of a catastrophic slide to up to 6°C of global warming. Unfortunately, when the nations met in December that year, they could not agree legally binding targets. As the months tick by and the 2015 deadline draws closer, we're still waiting.

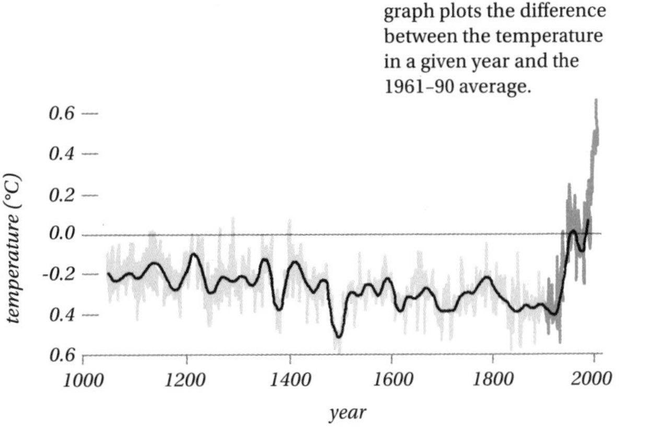

The so-called 'Hockey Stick' graph shows the change in global average temperature over the last millennium. The graph plots the difference between the temperature in a given year and the 1961–90 average.

22 How to build an Earth

- Fragments of a dead star
- Earth: the early years
- Shifting continents
- Creation and destruction
- What earthquakes tell us about our planet
- Everything is in motion

Charles Lyell, mentor to the great naturalist Charles Darwin, was one of the first to turn geology, the study of the Earth, into something scientific. But many of the big things we take for granted today, such as the age of our planet as more than 4.5 billion years old, were only established in the latter half of the 20th century. The idea that the surface of Earth is in flux, gradually making and destroying mountains and oceans over millions of years, is newer still.

Fragments of a dead star

Before there was an Earth, there was a chaotic cloud in place of our Solar System, containing the remnants of a dead star. Around 5 billion years ago, a massive star reached the end of its life and exploded, throwing its innards out into space. Everything in our Solar System today comes from the vast cloud of swirling dust after that supernova. We are the cooked atoms of billions of years of stellar evolution.

Because of the energy of that supernova explosion, the dust cloud was hot, so the atoms and molecules and larger bits of dust moved rapidly, bumping into each other and eventually coalescing in certain areas. One of these areas became so dense that the atoms began to fuse and the gas cloud turned into a star, our Sun. The remaining dust continued spinning and orbiting the new star, trapped by its immense gravity. Gravity also brought together inert dust particles over millions of years. First these dust grains made small lumps, the lumps joined together to form boulders and the boulders eventually fused into the embryonic beginnings of a planet. This is how Earth was made.

Earth: the early years

Our planet's baby years were nothing like the inviting place we might recognize today. It was a nightmare world, continually being bombarded by other rocks and asteroids that had coalesced from the dust. One such collision with a huge asteroid (or perhaps even a small planet) knocked a huge chunk out of our planet. This chunk flew off into space but couldn't quite escape Earth's gravity field, settling into an orbit instead. When humans looked up to see this object billions of years later, they would name our planet's closest companion the Moon.

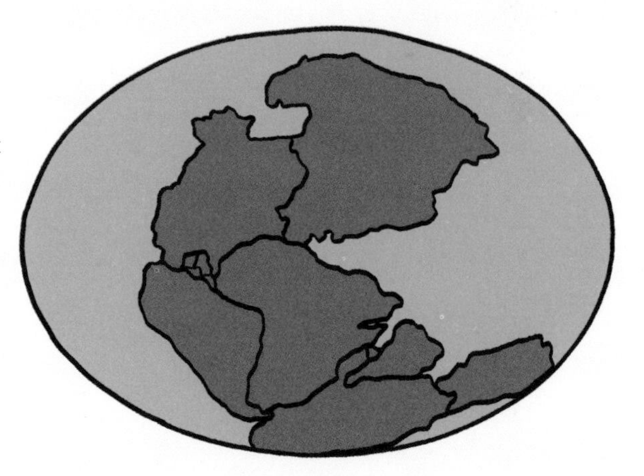

Around 250 million years ago, all Earth's continents were joined together as one giant land mass known as Pangaea (Greek for 'entire Earth').

The creation of the Moon had several long-lasting effects. The first was to knock the axis of Earth off-kilter, giving rise to the seasons as different surfaces of the planet are tilted towards or away from the Sun during its orbit. Tilt away from the Sun and we are in winter, while summer arrives when Earth is tilted towards the Sun. The second effect of the Moon was to stabilize the orbit of our planet, making it possible for the emergence of the delicate conditions needed to start life a billion years later.

Thanks to continued asteroid bombardment and radioactive decay of the heaviest elements swimming around on the molten Earth, the surface temperature kept on rising. The elements were mixed evenly throughout the globe and none of the structure and layers we know of today were evident. There were no oceans or even an atmosphere. Eventually, after the surface temperature had risen to around 2,000°C (3,600°F), the molten iron began to sink into the core. Earth eventually began to cool and separate into the structure we are familiar with today – a molten iron core surrounded by a mantle composed of liquid rock and topped off with a thin layer of crust.

The atmosphere and oceans, the things that eventually enabled life, were the combined result of volcanic activity releasing gases onto the surface of the planet and asteroids smashing into Earth. Most of the water in our planet's oceans came from icy rocks hurtling into Earth around 4 billion years ago, a period known as the 'heavy bombardment'.

The arrows show the different types of boundary between Earth's tectonic plates. Transform faults slide past one another, divergent faults (arrows in opposite directions) occur where plates slide apart, while convergent faults (arrows pointing towards each other) occur where one fault is subducted underneath another.

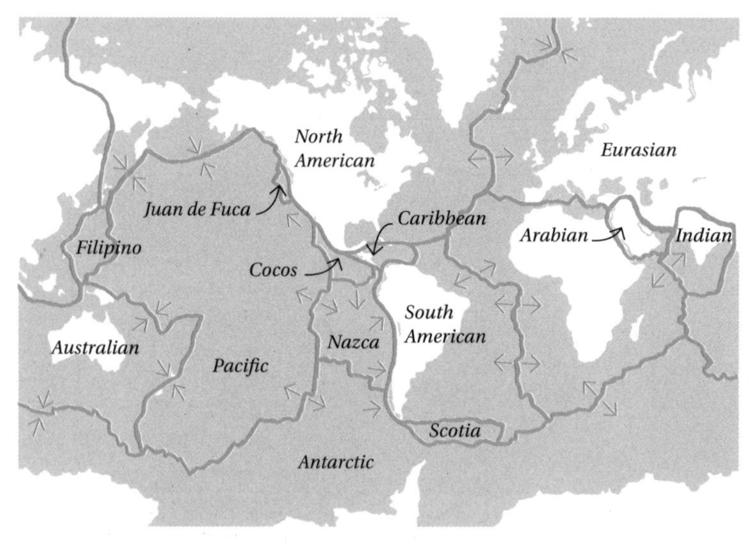

Life started around a billion years after Earth came into existence and, for another billion years, remained small and unimpressive because there was so little oxygen in the atmosphere. During this time, however, micro-organisms were subtly changing the composition of the air thanks to their various metabolic reactions until, eventually, there was enough oxygen in the atmosphere for complex life to evolve.

'Earth is on the move and always has been. There is movement on every timescale; the planet is like a symphony scored with reference to a metronome clocking everything from milliseconds to eternity.'

Richard Fortey

Shifting continents

The idea that Earth might be made of layers was not immediately obvious to early scientists. Edmund Halley thought Earth was a hollow shell with a crust several hundreds of miles thick. The novelist Jules Verne imagined vast caverns underground. Modern geology tells us that Earth's crust is a hard surface covering a molten, moving, mantle that is thousands of kilometres thick. Underneath that, at the centre of our planet, it is pure iron. Because of the intense pressure, it sits there, molten, and is the source of Earth's magnetic field. This field protects us from much of the harmful cosmic rays that bombard our planet.

The surface of Earth certainly seems steady and solid. But the liquid mantle 30 km (18 miles) underneath is continually driving an imperceptible motion. The Atlantic Ocean is widening, for example, at a few millimetres per year. The already enormous mountains of the Himalayan range are rising as India moves further into Asia. Deep under the surface of the Pacific Ocean, the seabed is being destroyed at an astonishing rate. All of these processes are driven by something that has shaped and reshaped the surface of our planet countless times in its four-billion-year history. Plate tectonics is the reason the continents take the shape they do today, and why they are so different from the supercontinent of Pangaea (Greek for 'all land') that existed at the time of the dinosaurs, when nearly all of Earth's landmass sat together in one large continent. Over billions of years, plate tectonics alters the shape of continents, oceans and, yes, countries.

The seeds of this idea started in the 17th century, when philosophers couldn't help but notice the uncanny fit between the west coast of Africa and the east coast of South America. But, obvious as it might be from hindsight, plate tectonics only took off as the established idea in the 1960s. The current model for the surface of Earth is that it is made up of 12 enormous plates that move around on the currents in the sea of molten rock in the mantle below. The plates are in constant motion and, most recently with Pangaea around 200 million years ago, sometimes come together to form a massive supercontinent.

Creation and destruction

All over the world, bits of the crust are being destroyed, and new areas of land are being created. Where magma pushes up through the crust, the tectonic plate can be torn apart and, if water fills the gap, the process forms oceans – the Rift Valley in east Africa is a good example, as is the growth of the Atlantic Ocean. Oceans also shrink. When continental plates meet and push up against oceanic plates, the latter might sink underneath the former, creating a subduction zone. At these places, and there are several in the Pacific Ocean, the crust that sinks melts into the mantle below, leaving behind its solid form. Between the upswelling magma and the subduction, a natural cycle determines what grows and what sinks.

The collision of continents causes a thickening of the crust that we see on the surface as mountains. When India smashed into Asia more than 35 million years ago, it began the Himalayan mountain range. Above we see Everest rising almost 9 km (5.6 miles) from sea level. But underneath is something more impressive – where normal crust is around 30 km (18 miles) thick, under Everest it is more than double that. The Alps are the result of the collision between Europe and Africa, and the Mediterranean sea is a temporary, water-filled, gap created by the plates. Volcanoes are produced by a different process. Surges of magma from the mantle pierce the crust at particular places (called hotspots) and the sudden outpouring of rock emerges as a volcano. As a tectonic plate slides over a hotspot, the result is a chain of islands (such as the Philippines) or mountainous volcanoes. The islands of Hawaii were produced like this – the biggest Hawaiian volcano, Mauna Kea, is several kilometres higher than Everest from base to tip.

What earthquakes tell us about our planet

Sometimes, when plates try to slide past each other or subduct under one another, they do not move with ease. Perhaps because of friction neither moves as nature intended. Instead the plates stay in one position, building up enormous energy in the form of tension at the plate boundary. This tension is the root of earthquakes.

At some point, the tension at the plate boundary will be too much for any frictional force to keep things in place. The vast store of energy will be released suddenly, causing the ground above to open up and shake. The effects can be devastating, with hundreds of aftershocks in the days or weeks after the first quake. Any or all of these can cause direct destruction, such as the collapse of buildings, or lead to landslides or tsunamis. Even if the earthquake happens somewhere far from population centres, the effects can be catastrophic. One of most destructive earthquakes ever recorded occurred in a subduction zone in the Indian Ocean near Sumatra on Boxing Day 2004, when a quake of moment magnitude 9.2 triggered tsunamis that killed more than 200,000 people.

Earthquakes occur at three main types of fault. At a strike-slip fault, the two sides of the fault slide past each other horizontally. At normal and thrust faults, the movement is primarily vertical. Normal faults occur at divergent boundaries where the crust is being extended. Thrust faults occur where the crust is being shortened at convergent boundaries.

Most earthquakes are, thankfully, nowhere near as devastating as the 2004 event. The vast majority, in fact, are not even strong enough for people to register them – around half a million earthquakes of different sizes occur in the world every year, and only a fifth of them are strong enough to be felt. Earthquakes release their energy in the form of seismic waves, and the way these waves travel through the planet are crucial to giving scientists their understanding of Earth's internal structure. The waves travel at different speeds through different materials, and by looking at how they refract and reflect at the boundaries between Earth's changing structure, geologists can build up an image of the planet's interior.

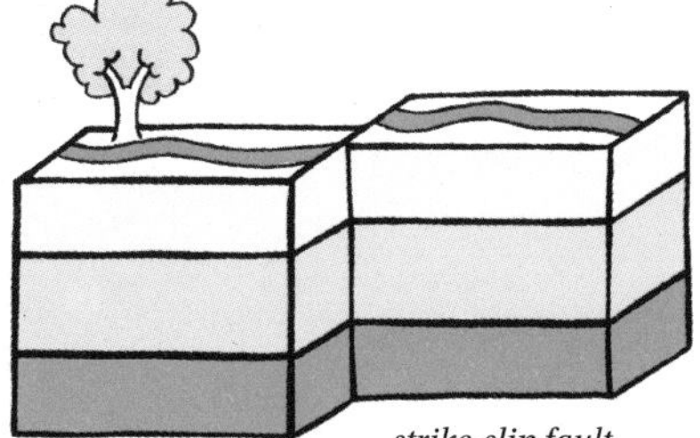

strike-slip fault

Further hints of what lies underneath come from volcanoes: the lava they throw out comes from deep under the crust, giving us a glimpse of what the mantle might be like – though no-one is sure if lava is a direct sample of mantle as a whole or if it comes from only the uppermost layers, leaving the rest of the liquid rock still unobserved. Geologists simulate the extreme temperatures and pressures they think occur inside Earth using computer models. It is safe to say that none of the information has come from digging. The deepest hole ever dug, in Russia, is barely 12.8 km (8 miles) deep, nowhere near enough to get through the crust.

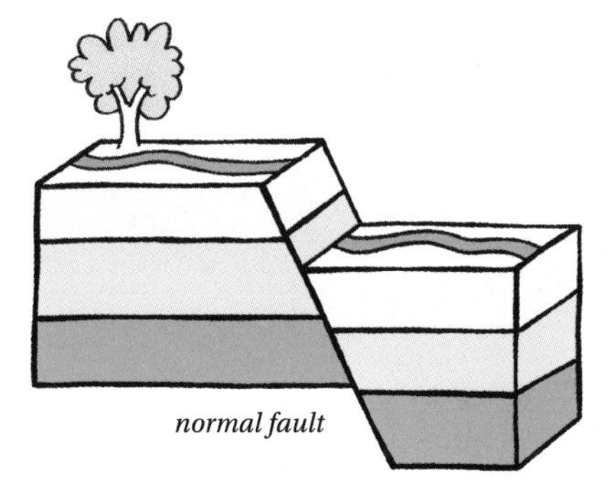

normal fault

Everything is in motion

Nothing on Earth's surface is permanent. In a few million years from now, says British palaeontologist Richard Fortey, the Himalayas will be rounded hills, Hawaii will have sunk from view and the Pacific Ocean will have been swallowed back into the mantle. Long after humans have died out and the climate has turned into something even more inhospitable than it is today, Earth will continue to change. 'Earth is on the move and always has been,' he says. 'There is movement on every timescale; the planet is like a symphony scored with reference to a metronome clocking everything from milliseconds to eternity.'

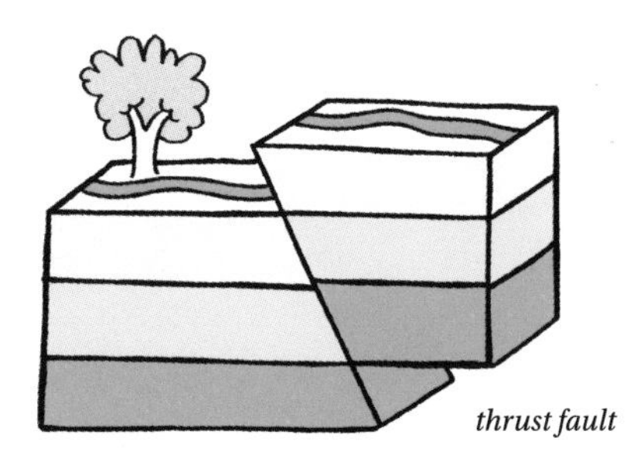

thrust fault

23 How to control the weather

- How Earth makes weather
- Wind
- How we use weather
- Devastation
- Controlling the skies
- Cloud seeding
- Storm prevention
- Future ideas
- If we could change the weather, should we?

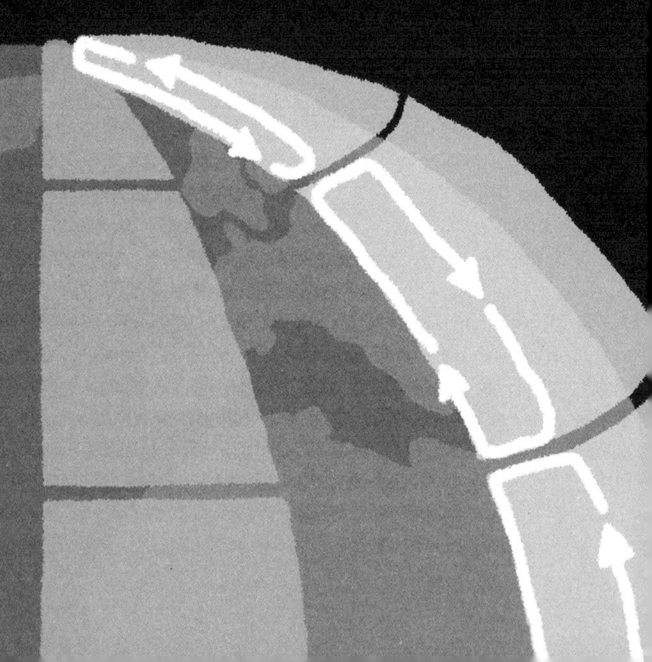

Shamans and storytellers have long linked human fate to the weather. To control it has been a power akin to being a god. Native Americans dance to summon the rains; in Homer's* Odyssey, *the eponymous hero's fleet is blown off course after his crew misuse the four winds gifted by Aeolus, keeper of winds. Today, our lives are no less reliant on the temperature of the air or the humidity. Weather forecasts help us with everything from what we plan to do or wear tomorrow, what is possible in the world and even how we might feel.

How Earth makes weather

'Conversation about the weather is the last refuge of the unimaginative.'

Oscar Wilde

All the rain, wind and temperature changes that we know of as weather are the result of the chaotic interactions between the sun, sea and air. The day-to-day weather that concerns us takes place in the lowest layer of the atmosphere, called the troposphere, which is up to 20 km (12 miles) thick depending on where you are, and contains three-quarters of the whole atmosphere's mass. It also contains almost all of its water vapour.

Everything from a warm summer day to a wet, destructive storm begins with the input of energy from the Sun. The wide variation in weather around the globe comes from the fact that the Sun doesn't shine across the world uniformly and it only ever shines on half of the world at once. At the equator, sunlight comes in vertically from above, whereas in polar latitudes, it comes in at a shallower angle, and is weaker because it has had to travel through more air. Clouds absorb around a quarter of the incoming energy, the atmosphere absorbs another quarter and the rest travels to the surface. Oceans store huge amounts of the sun's energy, while continents tend to reflect most of it back into the air as heat, which explains why a village on a landlocked savannah is usually so much hotter than a coastal town.

Oceans are vitally important in carrying energy all over the globe, transferring warmth from equatorial regions to the colder polar latitudes via huge water currents that crisscross the world like free-flowing expressways. Warm water, for example, snakes its way up from the Gulf of Mexico along the eastern coast of North America, on to the UK and then towards Iceland. Known as the Gulf Stream, this surface current gives the UK a much warmer climate than it should really have, given its northerly latitude. It warms Florida and gives the residents of Scotland and Norway much more tolerable weather compared to their counterparts on similar latitudes in Canada and Russia. By the time the Gulf Stream reaches Europe, the water has cooled because of surface winds, which evaporate away some of the water, making

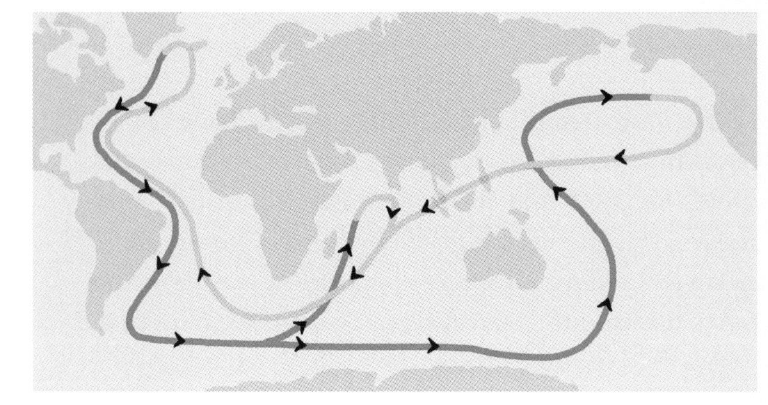

The movement of water from ocean to ocean is known as the thermohaline circulation. The dark blue lines are cold, salty deepwater currents, while the light blue lines are warm surface currents.

the current saltier and more dense. It therefore sinks to the bottom of the ocean and enters the North Atlantic deep water current, a languid stream of water that emerges eventually in the Southern Ocean after more than 1,000 years.

Wind

The atmosphere also transfers heat from one place to another, albeit far more quickly than the oceans. When air warms up, either because of direct sunshine or because of heat emitted by the ground below, the molecules move faster and push each other apart. When it cools down, the opposite happens and the air becomes more dense. Wind is the result of air molecules travelling between these differentials in air pressure at different places. If you travel from the equator to the pole in a straight line, you cut across three distinct sets of atmospheric cells arranged like belts around Earth. These looped air currents take heat from warm areas to cold. At the equator, the convection loops are known as Hadley cells: warm, moist air rises in a low-pressure area at the equator, travels towards the pole and, at around 30 degrees latitude, drops back to the surface in a high-pressure area. The next belt of cells moving north or south are known as Ferrel cells, and then finally polar cells sit furthest from the equator in both hemispheres. These enormous cells, and the flows of energy through and between them, give rise to the temperature differences, precipitation and wind that we experience.

How we use weather

Hadley cells are also responsible for the trade winds, a prevailing pattern of easterly surface winds that blow across Earth's tropical regions. They filled the sails of ships of old, enabling Europeans to reach the New World and establish trade routes across the oceans. And, higher up in the atmosphere, where the weather-filled troposphere meets the stratosphere, another vital set of winds cross the globe. If you've ever wondered why flights from New York to London are faster than going the other way, it is because your plane is being partly propelled by the jet stream, a point where two convection cells meet the boundary between atmospheric layers. Winds here have been measured at up to 400 km/h (250 mph) and, in the northern hemisphere, they hurtle along from west to east. And they often carry more than aeroplanes – storms can also be carried from one side of the ocean to the other.

Devastation

The mechanisms behind wind and rain are well understood. But predicting weather is still fraught with uncertainties, as anyone caught out in a shower of rain will attest. It is even more difficult to pin down freak occurrences, where surface temperatures and humidity happen to combine to create huge storms. Often this can lead to heavier versions of normal weather – faster winds, more rain or bigger hailstones. But sometimes the natural world also demonstrates just how much power it can summon up.

Hurricane Katrina was behind the deaths of more than 2,000 people in New Orleans in 2005 and caused tens of billions of dollars in damage. Hurricanes get their power from the seas – the warmer the sea-surface is, the more powerful a hurricane can become. Katrina began as a low-pressure weather system over the Bahamas, becoming a category 5 storm as it travelled across unusually warm waters in the Gulf of Mexico. When it made landfall south of New Orleans many days later, its intensity had actually reduced to a category 3 hurricane. It still carried sustained wind speeds of up to 200 km/h (125 mph).

Hurricanes, also known as typhoons or cyclones depending on where in the tropics they occur, form every year during a few months towards the end of summer, when sea-surface temperatures are high enough to power the storms. They start as low-pressure areas, sucking in surface wind and directing it upwards. This creates a tunnel of rising warm moist air – the relatively calm 'eye' of the storm – which is surrounded by a ring of thunderstorms that form as the moist air spreads out and cools.

Controlling the skies

Given the scales and energies involved, can we do anything to affect something so complex and so huge? Controlling the weather might seem a ridiculous idea, but there is a lot of work going on to do just that. The atmosphere is a chaotic system – even the tiniest difference in starting conditions can result in a very different outcome days or weeks later. Writing on whether it would ever be possible to control such a system, Ross Hoffman, head of research at US firm Atmospheric & Environmental Research Inc, made his thoughts clear. 'Extreme sensitivity to initial conditions implies that small perturbations to the atmosphere can effectively control the evolution of the atmosphere, if the atmosphere is observed and modelled sufficiently well.' It is, he was saying, all in the preparation.

In each hemisphere, there are three cells in which air circulates through the entire height of the atmosphere. The cells closest to the equator are called the Hadley cells. Next to these come the Ferrel cells, outside which are the polar cells.

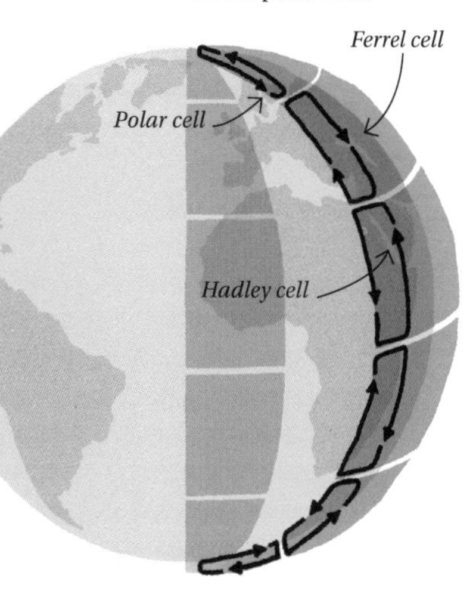

Earth's atmosphere is made up of five distinct layers. Most weather phenomena occur in the troposphere, which contains about three-quarters of the atmosphere's gases and 99 percent of its water vapour.

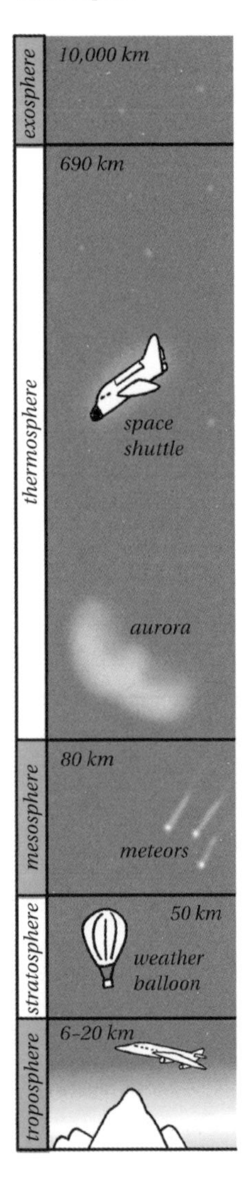

Cloud seeding

One type of weather modification has already proved somewhat successful. In November 2009, an unseasonal smattering of snow fell onto Beijing for 11 hours. According to Chinese authorities, the snow was the result of 186 rockets that had been fired into the sky, each one loaded with chemicals meant to 'seed' clouds and encourage precipitation. In June 2004, when rain was threatening to stop play, Paul McCartney's concert promoters arranged for three jets to dump dry ice into the skies over St Petersburg's Palace Square, where the former Beatle was due to play his 3,000th gig.

In the latter half of the 20th century, many others around the world, the US military in particular, tried out 'cloud seeding', which usually involves sending a fine spray of dry ice or silver iodide particles high into the atmosphere, providing water vapour up there with something to latch on to in order to form clouds. Cloud seeding does seem to work, but it is hard to prove that it ends up doing anything special – who is to say that the rain caused by a seeding exercise would not have fallen anyway a day later?

Another way to make it rain is to fire lasers into the sky. Using very short pulses of light, scientists at the Free University of Berlin and colleagues in Germany, Switzerland and France managed to ionize the air and start condensation high in the atmosphere.

Storm prevention

Silver iodide particles have also been used in an attempt to weaken storms. From 1962 to 1983, the US government ran a project called Stormfury. The idea was that the particles would cause the supercooled water in the storm to freeze, thereby disrupting the structure of the winds required to keep the storm going. The project ran several experiments in the Atlantic and, at first, seemed to be onto a good thing. But further research showed that there simply was not enough supercooled water in the storms for the silver iodide to have any significant effect. In addition, meteorologists saw that untreated hurricanes often underwent the same structural changes naturally that silver iodide treatment was meant to have enabled artificially.

Future ideas

The power of a hurricane comes from the surface temperature of the water. If you can reduce that, then it should be possible to take some of the strength out of a storm. The US company, Atmocean, has a potential solution in its 1,000 m (3,300 ft) long, 1.5 m (5 ft) wide plastic tubes. The tubes sit near the surface of the ocean, using the motion of the waves to pump cold water from the deep to the surface. If an array of these tubes were installed in hurricane hotspots (such as the Gulf of Mexico), they might be able to bring enough cold water to the surface of the sea to reduce the temperature by one or two degrees whenever a dangerous storm was imminent.

Another idea is to create an oil slick on the ocean in the path of the hurricane. This film would prevent water from evaporating from the sea, limiting how big the storm could become. There are problems with this approach, of course. During a storm, the sea will also churn, so the concept of a sea surface is somewhat moot. An oil slick would quickly dissipate in those conditions.

A third concept involves dropping tonnes of a water-absorbing powder into the air, which would extract water vapour from the air. Proposed by a US company called Dyn-O-Mat, it might stop a storm in its tracks by drying out clouds and then, as the cold particles fall onto the sea below, they would cool the surface, further weakening any storm that passes over it.

If we could change the weather, should we?

There is much to be sceptical about when it comes to the hubristic idea of wanting to control a weather system that might be hundreds of kilometres across and contain staggering amounts of kinetic energy. How could any human intervention possibly have an appreciable effect on something as big as a hurricane or typhoon? And what about unintended consequences? Cloud seeding sounds like a good idea if your bit of land needs extra water but it can hardly be targeted – what if your neighbouring country doesn't want their land rained on that day? The US, perhaps not surprisingly given its history of trying to change the weather, has laws regulating this sort of work. Some ideas require licences or public notification before they are carried out. In Oklahoma there is an exception for religious ceremonies, so Native Americans can perform rain dances without fearing any reprisals from the law.

Hoffman sounds a warning. 'The nation that controls its own weather will perforce control the weather of other nations. Weather "wars" are conceivable. An international treaty may be required to ensure that weather control technology be used for the good of all.' It might not be feasible to control hurricanes, typhoons and thunderstorms today but, if it is physically possible, someone somewhere will no doubt work out how to do it. Hoffman is probably not wrong in suggesting we stay prepared.

24 How to survive in space

- The effects of weightlessness
- Radiation death rays
- Cosmic dust and space junk
- Space shields
- What next for space exploration?

Humans have ventured a small distance from Earth and sent scores of probes to examine our Solar System. For such a curious species as ours, the first 50 years of the Space Age has produced the kind of knowledge that has only whetted our appetite. But the challenge of exploring further into space carries even more danger. The tiniest rocks and bits of debris can destroy a spacecraft, while radiation from faraway cosmological objects can destroy human cells and ravage our immune systems.

The effects of weightlessness

Space is a hostile environment for biological life. Never mind the lack of air to breathe (at least that can be solved), it's the lack of gravity that can have some of the most initially unpleasant effects for anyone heading into orbit around Earth. The human body needs gravity to determine which way is up and which is down. If there's any confusion, the body will respond in the only way it knows how – by inducing nausea and vomiting. For astronauts, 'space adaptation syndrome' is a natural reaction to the weightless conditions inside the space shuttle or International Space Station (ISS) as it orbits Earth. Motion-sickness drugs can help, and the body eventually does get used to the situation after a dew days. Weightlessness also affects an astronaut's muscles, which can become weak through lack of use. Bones can start to weaken and lose minerals as they are not feeling the effects of the gravity that normally keeps them strong. Both these effects can be minimized by exercise and diet in space – the ISS contains exercise bikes and treadmills to keep astronauts healthy against the wasting effects of weightlessness.

Astronauts returning from extended missions in low-Earth orbit need time after they get back for their bodies to return to optimum functioning. The return to normal weight, for example, can cause an inability to maintain proper blood pressure when standing up. The time needed for astronauts to re-adapt to Earth goes up depending on how long they have been in space. This is fine for astronauts returning to Earth but could be problematic for trips to Mars, where a crew might be in weightless conditions for a year or more on the trip to the red planet.

Radiation death rays

Unseen and far more dangerous things lurk in the radiation and energetic particles that zoom through empty space. Earth's atmosphere protects the surface from most of the harmful effects of these so-called 'cosmic rays' – spewed out by our Sun and other stars or exotic objects in our galaxy and

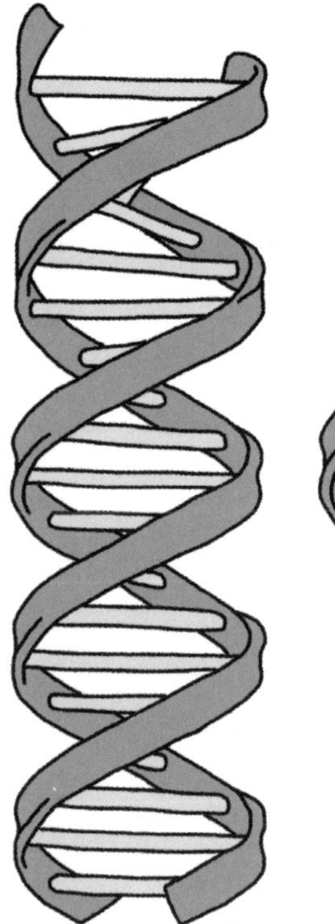

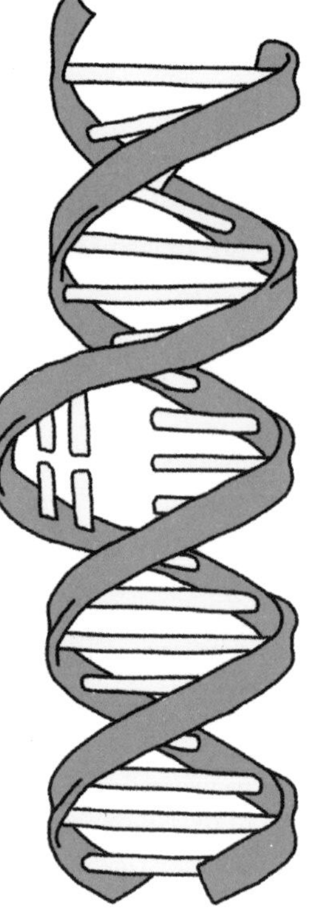

beyond. Though known as rays, they are almost entirely streams of individual protons, along with some helium nuclei and even tinier proportions of other elements. Each particle bears a colossal energy, many orders of magnitude more than anything produced even in the biggest particle accelerators on Earth.

Without our atmospheric shield, our fragile biology would be at the mercy of cosmic rays. The energetic positively charged particles can pass through skin and tear apart the molecules, including DNA, in our cells. DNA carries our genetic blueprints, the instructions on how to grow, how to regulate, which hormones to produce and when. Ionizing radiation, such as cosmic rays, can cause mutations in DNA that can either kill a cell or, worse, lead it to become cancerous.

Ionizing radiation can break strands of DNA, which can cause a change in the structure of the molecule. If this is not fixed, this damage can turn into cancer. The DNA on the right has been damaged by radiation.

Scientists have estimated that, on a three-year trip to Mars, almost a third of an astronaut's cells would be hit by cosmic rays containing fast, heavy particles containing lots of protons, while virtually all their cells would be hit by the more numerous lighter particles. The heavier a cosmic ray particle, the more damage it can do and each hit raises the likelihood of an astronaut sustaining the type of damage that could lead to cancer or death.

Cosmic dust and space junk

As well as the concerns about high-energy particles, there are worries about physical objects in space – everything from small dust grains in interstellar space to rocks and asteroids that could collide with a spacecraft, blowing a hole in the hull. The dangers can be man-made too – the debris of half a century of launches into space sits in a shell around Earth. This vast flotilla of stuff includes screws, bolts, flecks of spacecraft paint, tools, lens caps, dead satellites and bits of frozen water or coolant. All of these are spinning around Earth at more than 28,000 km/h (17,500 mph), so even the tiniest piece can tear a hole through the metal hull of spacecraft.

Space shields

The potential dangers in space are well-recognized by space agencies, though there is much work to do in quantifying the effects that radiation or physical objects could have on humans or spaceships. In 2003, NASA set up the Space Radiation Shielding Program at the Marshall Space Flight Center in Huntsville, Alabama, to coordinate work in the area and find ways to protect anyone or anything on long space missions. So far, scientists have proposed three ideas for how to protect against the dangers of space travel: building a solid shield around the spacecraft, using magnetic fields, and electrically charging the hull of a spacecraft to repel cosmic rays.

The first idea, building a solid shield, mimics Earth's atmosphere. This would keep out both physical objects and cosmic rays and, in principle, is technically quite simple. Eugene N. Parker, emeritus physics professor at the University of Chicago and the world's leading expert on interplanetary gas and magnetic fields, says mimicking the shielding effect of the atmosphere would require around 500 grams (1 lb) of material per square centimetre. This is roughly the weight of a column of air, with a square centimetre cross-section, at the peak of a 5,500m (18,000ft) mountain. In an article for *Scientific American*, Parker wrote: 'If the material is water, it has to be five meters deep. So a spherical water tank encasing a small capsule would have a mass of about 500 tons. Larger, more comfortable living quarters would require even more.'

A magnetic shield would protect a spacecraft by deflecting incoming cosmic rays.

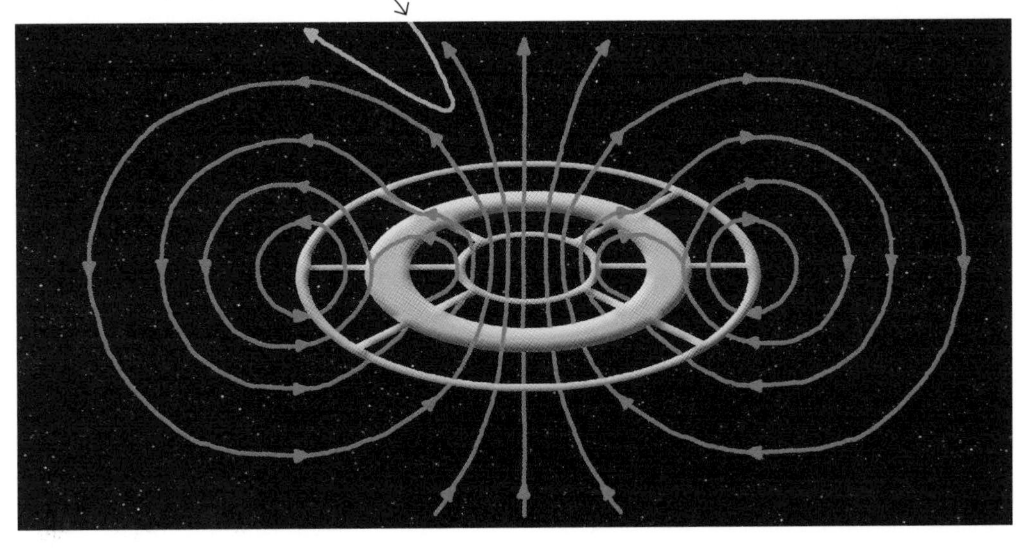

Water has lots of positively charged hydrogen ions (useful for repelling positively charged cosmic rays), and is also a good idea because it would be needed by the astronauts anyway, for obvious reasons. Another material that could be used for a shield, says Parker, is ethylene. This is even richer than water in hydrogen and has the added advantage that it can easily be turned into a solid, avoiding the need for a tank to contain it and also negating the chances of any leakage. 'Even so, the required mass would be at least 400 tons, still not feasible,' says Parker.

What about a magnetic shield? If an electrically charged particle, such as those in cosmic rays, were to encounter a magnetic shield, it would be deflected so that it moves off at right angles to the initial direction of motion. Depending on the set-up of the magnetic shield around a spacecraft, in-coming cosmic rays can be bounced off in almost any direction.

The problem with this promising idea lies in engineering. Cosmic ray particles move very fast, at significant proportions of the speed of light, and their kinetic energy is therefore very high. Parker says that, to stop them within a reasonable distance around a spacecraft would require the field to be around 20 teslas. This is 600,000 times stronger than the strength of the magnetic field at Earth's surface, and means building a magnet with superconducting wires, rather like the ones used in modern particle accelerators. The costs of such a device would be astronomical.

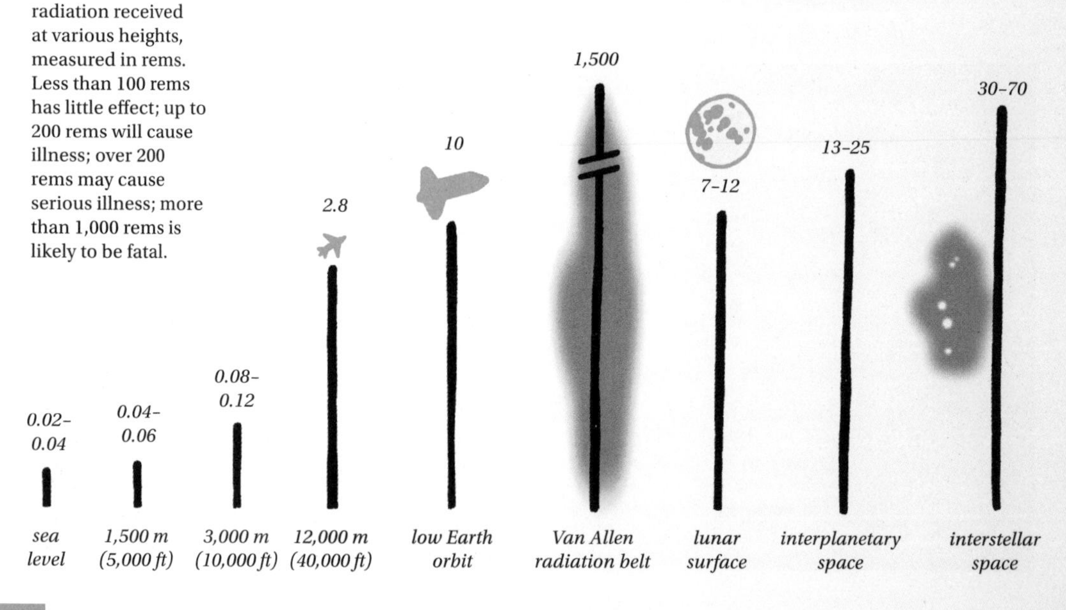

The amount of radiation received at various heights, measured in rems. Less than 100 rems has little effect; up to 200 rems will cause illness; over 200 rems may cause serious illness; more than 1,000 rems is likely to be fatal.

Another issue is that no-one is quite sure what the biological effect of such a strong magnetic field might be on the human body. To give us some idea, Parker recalls how a colleague at the University of Chicago placed his head inside a 0.5 tesla magnetic field: 'Any motion of his head produced tiny flashes of light in his eyes and an acid taste in his mouth, presumably caused by electrolysis in his saliva.'

'Exploration is an important survival strategy in evolution.'

RICHARD B SETLOW

The third option is to positively charge the hull of a spacecraft. Given that cosmic-ray particles are also positively charged, they would be repelled by the positive electrical field around the craft. To make this work, the outer skin of the spacecraft would have to be charged up to around 2 billion electron volts relative to the space surrounding it.

Again, electrical shielding comes with a whole list of problems attached. The biggest one is the unintended consequence of the electrical field on particles that are not cosmic rays. Parker says that any electrons in the space thousands of kilometres around the ship (and there are a few electrons in every cubic centimetre even in 'empty' space) would be attracted to the hull, slamming into it with as much energy as any of the cosmic ray particles that the shield is meant to protect against. 'The electrons would produce gamma rays on impact with the spacecraft, and the intensity of that bombardment would be staggering, dwarfing the original problem,' he says. The energy required to charge up the entire spacecraft to 2 billion electron volts in the first place is also mind-boggling, equivalent to the entire electrical requirement of a small town.

What next for space exploration?

'It would be too bad if the romance of human space travel ended ignominiously with cosmic rays making it infeasible,' says Parker. 'Capable people might be willing to go to the Moon or Mars just for the adventure, come what may. Even so, the radiation hazard would take the lustre off the idea of human space travel, let alone full-scale colonization.'

Trying to maintain our biological stuff, which has evolved over millions of years to survive on Earth, in the harsh environment of space is a head-scratchingly grave limiting factor. If we cannot find an economical solution, perhaps the answer is to forget sending ourselves into the cosmos and to send robotic probes armed with sophisticated sensors instead. We could then enjoy a virtual reality tour of faraway planets and star systems from the comfort of an armchair. Though what Columbus might think of such a distant method of exploration is anybody's guess.

25 How to find the missing parts of the Universe

- The Coma conundrum
- How much is there?
- What could it be?
- The other dark force
- How might these missing bits be detected?

There is something wrong at the heart of cosmology, a problem that should spook anyone with any scientific interest in knowing why we exist: most of the Universe is missing. After hundreds of years of star gazing, billions spent on studying the tiniest particles and sending probes into space, and scores of Nobel prizes awarded on work that illuminates the fundamentals of the cosmos, we still don't have any idea what most of the Universe is made of.

The Coma conundrum

Swiss astronomer Fritz Zwicky had inklings that there was something strange about the Universe in the 1930s when he was studying galaxies in the Coma cluster. He found that galaxies at the edge of the cluster moved faster than expected and, given how much material he could see in the galaxy cluster in the form of stars, the gravitational force should not have been sufficient to stop the galaxies from breaking free from the cluster. He wondered if there was greater mass in the cluster than he could see. Something non-luminous, perhaps, that was present in greater quantities than the stars and dust, but which his instruments could not detect. A higher mass in the cluster would explain the anomalous orbital velocities of the outer members of the galaxy cluster, allowing them to move at the speeds he observed and remain bound to the whole.

In the following decades, other scientists corroborated Zwicky's findings. Astronomers now know that the mass of galaxy clusters must be bigger than the total mass of the individual galaxies in them, while galaxies themselves are more massive than the constituent stars and dust clouds observed by ground- and space-based telescopes. These instruments have shown that dark matter is something exotic and beyond our experience – if any of it was composed of the atoms and particles we are familiar with, it would have left fingerprints detectable by the increasingly sensitive experiments set up to monitor the cosmos. Dark matter's importance in creating the Universe as we see it today has also grown stronger. Today, cosmologists believe that it is the scaffold that holds together everything from galaxies to galaxy superclusters. Clumps of dark matter give structure to the Universe, the nuclei around which clusters of normal matter (in the form of stars and galaxies) can crystallize.

How much is there?

By indirect measurements to account for the speeds of stars and galaxies at the edges of clusters, astronomers have calculated that dark matter must make up a quarter of the mass of the Universe. Much like normal matter, it is

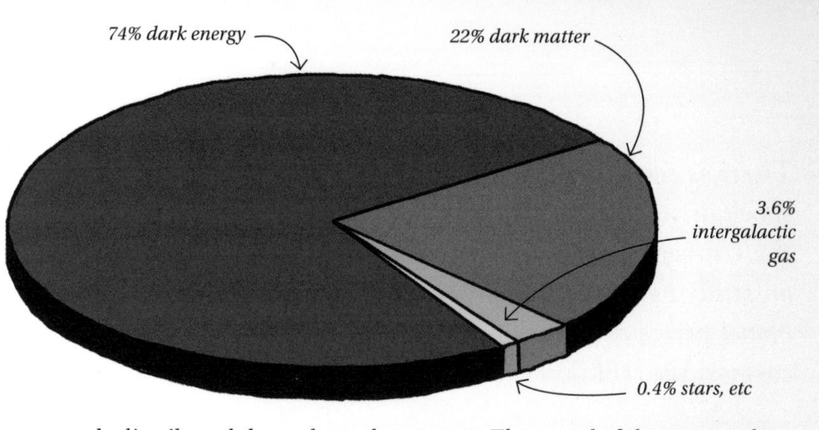

Nearly three-quarters of the mass of the Universe is made up of dark energy. Visible matter comprises just 0.4 percent of the total.

not evenly distributed throughout the cosmos. The speed of the stars at the edge of the Milky Way indicate that our galaxy is embedded in a halo of dark matter a thousand times bigger than the galaxy itself and at an average density of about one third of a proton mass per cubic centimetre of space. This is more than 100 times the average mass density of the rest of the Universe. Astronomers have also worked out that dark matter can be measured in 'bricks' that are a thousand light years across and, if you could measure their temperature, which you can't because they do not radiate, it would be around 10,000 kelvin. There are also local variations depending on how objects move around – the dark matter in Earth's vicinity changes by about 10 percent throughout the year as we orbit the Sun, for example, as our planet either moves in the same direction or away from the continual tidal wind of dark matter particles coming from the galactic halo.

What could it be?

It's easier to say what dark matter probably isn't. The initial prime suspects were neutrinos left over from the Big Bang, tiny particles that are abundant and hardly ever interact with normal matter. Even with a tiny mass, could that add up to the missing mass in the Universe? A quick calculation shows this cannot be true, however, as even the huge numbers of neutrinos we know about cannot account for the amount of dark matter out there. Successive other explanations based around known particles of matter have also fallen by the wayside: dead stars, black holes and clusters of inert dust or planets. Beyond the fact that there is lots of it and that it feels the effects of gravity, all we know about dark matter is that it does not seem to interact with the electromagnetic force, which means it doesn't shine. Dark matter cannot be made, therefore, from any of the components of the current Standard Model of physics, which is the quantum mechanical description of the matter and force particles that are known to exist and which we interact with. This is made up of all the quarks (constituents of the proton and neutron), leptons (including the electron and neutrino) and force-carrying particles (such as the photon). None of these particles fit the descriptions attributed to dark matter.

This means physicists are on the hunt for entirely new subatomic particles, outside the Standard Model, which can only be created at high energies in the most powerful particle accelerators. This hunt has led to a joining of minds between physicists who study the very biggest things in the Universe and physicists studying the very smallest constituents of matter and force. To help with the search, cosmologists made friends with their colleagues down the corridor in the particle physics laboratories. Both are eagerly awaiting the results of collisions at the Large Hadron Collier (LHC) in Geneva, Switzerland. This particle accelerator will smash together protons at energies that have not been seen in the Universe since the moments after the Big Bang. In the process, the protons will be torn apart and, physicists hope, as-yet-unseen particles could be created. Some of these particles might well fit the bill for dark matter, some cosmologists hope.

Candidate dark matter particles are collectively known as 'weakly-interacting massive particles' (WIMPs). Like neutrinos, they would leave little or no trace on normal matter but, unlike neutrinos, could be up to 100 times more massive than a proton.

One promising source of WIMPs comes from a theory called supersymmetry. This says that every matter and force particle in the Standard Model has a twin that we have yet to find. These 'superparticle' twins are heavier than their more familiar counterparts and have not existed naturally since the first 10^{-35} seconds after the Big Bang. But they might be produced as a result of the enormous energies that will be achieved at the LHC. A leading dark matter candidate is the neutralino, a particle that is the supersymmetric twin of the more familiar neutrinos left over after the Big Bang. Another possibility is the axion, an elementary particle thought up by theoretical physicists in recent decades to explain glitches in the strong nuclear force, which normally binds quarks together in the proton and neutron. Could this (or its supersymmetric partner, the axino) be present in enormous numbers around galaxies and stars?

This image was taken by the Hubble space telescope. The clusters of galaxies circled contain particularly large amounts of dark matter, which shows up on the image as the lighter areas.

The other dark force

The mystery surrounding dark matter is as nothing compared to dark energy, a mysterious force (or substance) that

'What is there in places empty of matter? And whence is it that the Sun and planets gravitate toward one another without dense matter between them? Whence is it that Nature doth nothing in vain?'

Isaac Newton

accounts for everything in the Universe that is not dark or normal matter. Cosmologists know it must exist because the Universe is expanding faster than they thought it should be. That the Universe was expanding had been known for most of the 20th century but, by now, the galaxies should be slowing down due to the mutual gravitational attraction of observed (and, thanks to dark matter, unobserved) matter. Some theories had even suggested the Universe should have stopped expanding altogether at some point and then started imploding, leading to a big crunch some time in the future – a reverse Big Bang.

But in the 1990s, observations of distant supernovae showed that conventional thinking was wrong. The Universe was not just expanding, it was expanding faster than ever. Given all the matter in the Universe, something must be working against the tug of gravity. To explain it, they postulated the existence of dark energy. But what this stuff is, is anybody's guess. Unlike dark matter, there are not many viable theoretical candidates. To understand it, physicists have turned to some of the less-visited bits of quantum theory. The equations allow, for example, for odd behaviour in some subatomic particles that give rise to fluids with some of the properties ascribed to dark energy.

A fluid with a negative pressure – hard to imagine, for sure, but possible under some specific quantum mechanical conditions – would give rise to a repulsive form of gravity. Dark energy, then, could be a fluid of particles in this bizarre state. Well, bizarre to us, but clearly commonplace in most of the Universe. Another idea is that dark energy could be some undiscovered fifth fundamental force of the Universe (the others being gravity, electromagnetism and the strong and weak nuclear interactions) that only manifests itself over cosmological distances.

How might these missing bits be detected?

Given its propensity to keep itself to itself, detecting dark matter is going to be difficult. But, never quick to give up, scientists have come up with a variety of detectors and probes designed to spot the faintest hints of anything unusual, which could at least provide clues or circumstantial evidence for the existence or properties of dark matter.

At the LHC, high-energy collisions will make new particles and some of these could have the properties ascribed to the WIMPs. Of course, being so undetectable, the WIMPs would not show up directly in the experiments, but rather as missing mass when the physicists piece together the mass and energy present in the accelerator before and after each collision.

One class of experiments relies on the fact that dark matter particles are streaming through Earth all the time. If one of them could be made to interact with a normal bit of matter, the collision would send out a flash of protons and electrons that sensitive modern instruments could detect. Buried under the Gran Sasso mountains in Italy is one such experiment: Xenon10. Here, 15 kg (33 lb) of liquid xenon sits under hundreds of metres of rock. Calculations suggest that almost a million-billion dark-matter particles should pass through the detector every week. With such a dense arrangement of xenon atoms, the hope is that just a tiny fraction of dark matter particles might hit something, releasing a flash of photons and electrons.

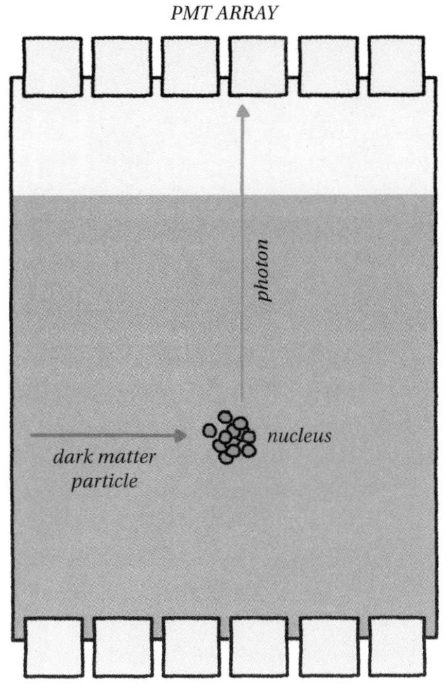

The Xenon10, buried deep under mountains in Italy, has been built to detect interactions of xenon nuclei with dark matter. The interaction causes the emission of a photon of light, which is detected by arrays of photomultipliers (PMTs).

Other detectors are on the hunt in space too. Neutralinos, the supersymmetric partners of the neutrino and another dark matter candidate act, as their own anti-particles. This means that, if two met, they would annihilate each other, releasing energy in the form of recognizable gamma rays. NASA's Fermi gamma-ray space telescope, launched in 2008, should be able to spot any tell-tale gamma radiation coming from the annihilations of neutralinos in the Milky Way's dark matter halo. Other experiments based on Earth's surface and on balloons are also on the lookout for these clues.

While there are lots of hypotheses and ideas for dark matter, dark energy is going to be a tougher nut to crack. The LHC might provide some clues for the structure of space or time that shed more light on the mysterious force, but no-one is sure that those measurements will do anything other than raise more questions. The hunt for this stuff is at its earliest stages and still largely theoretical. For the foreseeable future, dark energy might well remain resolutely in the dark.

26 How to programme your genes

- The Human Genome Project
- How genes work
- Information versus knowledge
- Here come the problems
- Epigenetics
- The future

In 2000, scientists published the first draft sequence of the human genome, the genetic blueprint for life. This century, we will use this knowledge to solve some of the puzzles of biology. Understanding the DNA in our cells will facilitate the treatment of almost everything that goes wrong with our bodies – from cancers to debilitating neurodegenerative diseases. Treatments may be found to improve the performance of athletes or students. We will be able to programme our genes and take charge of our molecular fate.

The Human Genome Project

In 1953, molecular biologists Francis Crick and James Watson published a paper in the journal *Nature* detailing the structure for the salt of deoxyribose nucleic acid (DNA). This discovery marked the start of the modern science of genetics. DNA, they said, is a molecule in the shape of a double-helix in which four nucleotide molecules (adenine, thymine, guanine and cytosine) form a specific sequence. The nucleotides bind together in pairs along the length of the DNA – adenine with thymine, guanine with cytosine – and are held by a backbone of phosphate and sugar molecules. Forty-seven years later, the multinational Human Genome Project finished sequencing the 3 billion nucleotide letters of the human genome.

John Sulston, the scientist who led the UK's side of the project, said that the knowledge of the human genome would have a far-reaching cultural effect on mankind, beyond its many practical uses. 'One should not underestimate how important this event is in human history. Over the decades and centuries to come this sequence will inform all of medicine, all of biology, and will lead us to a total understanding of not only human beings but all of life. Life is a unity, and by understanding one part you understand another.'

How genes work

A person's DNA, if laid out in a straight line, would stretch to the Sun and back several times. Its 3 billion nucleotide letters are contained on 46 long chains, called chromosomes, within the nucleus of each of our cells (except eggs and sperm, which have half that number). The chromosomes come in 23 pairs, one version from each parent. Studded along the chromosomes are the 24,000 genes that encode the thousands of proteins you need to construct a human. Before you think that's impressive, the number of genes an organism has doesn't mean it is more complex: a fruit fly has 13,600 genes and a yeast has 6,000; but thale cress, a common weed, has 26,000 genes.

Inside body cells a single-stranded version of DNA, called RNA, continually reads the genetic code in the nucleus and carries the information out into structures called ribosomes, which then build the proteins that are encoded by the DNA's instructions. Though every cell nucleus contains the entire genome, only parts of the genetic code are activated in different cells depending on the cell's function. Each gene is, on average, a few thousand base-pair letters long and contains codes for proteins to build muscle or cell walls, or make hormones or neurotransmitters that can regulate some aspect of behaviour. A person's genome is the full set of instructions – everything you could ever need to know – to make that person.

Genes account for less than 2 percent of the 3 billion letters of human DNA. Before scientists had properly grappled with the data from the human genome project, the remaining 98 percent of the DNA sequence was dismissively named 'junk DNA'. No-one knew why it was there or what it did. But the past decade has shown how this 'junk' DNA story might be wholly wrong. Careful analysis shows it to be made of sequences of genetic material that can copy themselves and jump to other parts of the genome. Other sections resemble parts of the genomes of other vertebrate animals, which suggests that they may be doing something useful, otherwise why would they be conserved through millions of years of evolution? Yet other parts of junk DNA really do seem like junk, with endlessly repeated letters.

The 'rungs' of the DNA ladder encode the genetic sequence, and each one is made from a pair of nucleotide molecules, also known as bases.

Look at someone sitting near you and you will see that they are probably very different from you. There's a good chance he or she will be a different height and have differently coloured hair, eyes and skin. Less obvious will be their different tolerance to types of headache tablet or their predisposition to specific diseases, such as cancer or stroke. Many of these physical properties are partially explained by the genetic cards we were dealt at birth – they are down to differences in the genetic code between you and that person.

This might make you think that, given the huge variation in how humans look, this difference would be reflected in our genetic sequences. In actual fact, you probably share at least 99.8% of your DNA sequence code with that very-different-looking person next to you. 'In every 1,000 of the genetic letters, there are two differences between people,' said Sulston when the genome sequence was announced in 2000. 'Most of the letters of the code are the same – but that still leaves an awful lot of differences. My way of looking at it is that we should take both morals from it. We should certainly regard ourselves as similar ... but we also have to respect our differences.'

Information versus knowledge

Discovering the sequence of nucleotide letters in a person's genome is just the start of trying to understand what it all means. The Human Genome Project gave scientists the basic letters and vocabulary with which nature writes its grand narratives of life. But the sequence is just the first step in understanding what those stories are telling us. Pinpointing what each gene does, how it varies between different people and how it can be 'fixed' if its mistakes end up causing disease is the work of the first few decades of the 21st century. It will be an enormous and involved task but, thanks to the rapid (and increasingly cheap) mechanization of biology, the results are already coming in thick and fast.

Sometimes a disease is caused by a single gene defect. The research into links between genes and disease in these situations has focused on studying a few genes from small numbers of samples, often from families with a higher than normal risk of suffering a particular illness. A classic example is cystic fibrosis, a hereditary condition that causes the airways in the lungs to become clogged with sticky mucus. Scientists pinpointed that it is caused when both copies of the CFTR gene do not work properly. However, many diseases are the result of the combined action of dozens or hundreds of genes. The approach taken for cystic fibrosis would have been too laborious even for the computerized biotechnology of the 21st century. The identification of genes really took off in 2007 with the introduction of a new technique called the genome-wide association study. Here, scientists look at DNA samples from thousands of patients for every disease, comparing them with control samples from healthy volunteers and looking at of genetic differences in each sample.

'Do these genes make me look fat?'

Labelling the human genome is crucial if we want to manipulate it. Say you want your child to be a star sprinter. You might look for genes that predispose your child to develop suitable muscles, such as ACTN3. One version of this gene is known to make a protein only found in fast-twitch muscles that help sprinters summon up explosive bursts of speed. A study of elite sprinters found that 95 percent of them had this explosive gene variant. If you could somehow select it, your kid could have an edge.

Epigenetics is the study of how genes behave in different situations, which ones are operating and which ones are turned off. This behaviour is controlled in two main ways: the addition of molecules called methyl groups (represented by 'Me' below) can suppress the activity of a gene; and the modification of the tails of proteins called histones, which alter the activity of the DNA strands wrapped around them.

Here come the problems

You might have guessed, given that there are not genetically modified athletes running around, that selecting for a behaviour as complex as sporting ability is not as simple as choosing the right version of ACTN3. Unlike cystic fibrosis, sporting prowess is governed by a multitude of genes that control everything from how your body lays down fat to how efficient it is at using oxygen. There could be hundreds, if not thousands, of genes that work together to create a sportsperson – and most of them are yet to be identified. In addition, access to training and number of hours sweating around a racetrack form a large part of who becomes an elite athlete. So selecting the correct ACTN3 is certainly no guarantee that your child will become an Olympian.

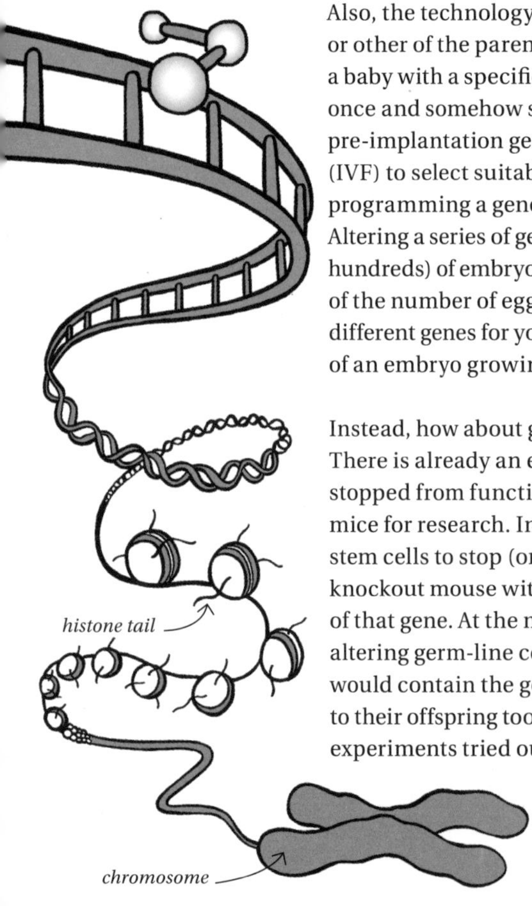

Also, the technology is messy and wasteful. Selecting a gene means that one or other of the parents has to have that gene to start with. And if you want a baby with a specific gene, you need to create several dozen embryos at once and somehow select those with suitable genes. This technique, called pre-implantation genetic diagnosis (PGD), is used with in-vitro fertilization (IVF) to select suitable embryos to be implanted into a womb. For programming a genome with specific genes, however, PGD is very clunky. Altering a series of genes at once would require the creation of scores (perhaps hundreds) of embryos and sorting through them to find a perfect match. Think of the number of eggs that would be wasted if you wanted to pick lots of different genes for your baby. Also, even if you found a perfect set, the chances of an embryo growing to full term in an IVF cycle are not 100 percent.

Instead, how about genetically altering a germ-line cell (an egg or sperm)? There is already an established technique whereby individual genes can be stopped from functioning and it is used to produce so-called 'knockout' mice for research. In these animals, artificial DNA is introduced into mouse stem cells to stop (or knock out) the action of a gene. Scientists compare the knockout mouse with a normal mouse, and thereby understand the function of that gene. At the moment it is illegal to do this with human stem cells, but altering germ-line cells would mean that any babies produced from them would contain the genetic change. And they would transmit the change to their offspring too. In any case, around 15 percent of the knockout experiments tried out on mice have proved lethal. Yet more produce disabilities in the animals. Many genes have multiple uses in humans – a gene that is associated both with a boost in IQ and a muscle condition could produce clever people in wheelchairs.

histone tail

chromosome

Epigenetics

There is a way to alter the behaviour of genes without resorting to germ-line cells or programming foetuses. All our cells contain the entire genome in their nucleus. But the whole genome is not active in every cell because many genes are silenced. Scientists have opened up a new world of 'epigenetics', defined as the changes in gene expression that are nothing to do with the sequence of DNA. This can change a huge range of things in an organism, from the shape of flowers to the colour of a fruit fly's eyes.

There are several mechanisms by which this can happen. One of them involves attaching molecules called methyl groups onto specific parts of a gene. In effect, the methyl group silences the action of the gene. The methyl group can be a temporary addition and can even be introduced via local environmental factors, including chemicals or food. Epigenetics is a fast-growing area of research as scientists map out an 'epigenome' to catalogue the various ways of affecting genes using environmental cues. Intriguingly, it even looks as though some epigenetic changes can be passed down through the generations, robbing DNA's claim to be the sole molecule of transmitting inheritance.

The future

Francis Collins, head of the US effort to sequence the human genome, admitted in an article published in *Nature* in 2010 that genomics still had a long way to go:

> *Some major advances have indeed been made: powerful new drugs have been developed for some cancers; genetic tests can predict whether people with breast cancer need chemotherapy; the major risk factors for macular degeneration [which causes visual impairment] have been identified; and drug response can be predicted accurately for more than a dozen drugs. But it is fair to say that the Human Genome Project has not yet directly affected the health care of most individuals.*

Technologies such as whole-genome sequences will come into their own as the cost per person falls below $1,000 in the next few years. This increasing mass of data will help identify even the rarest variations (or the 'dark matter' of the genome, as Collins describes it), and the prediction of diseases and likelihood of treating them with specific drugs will get better. 'Genomics has had an exceptionally powerful enabling role in biomedical advances over the past decade,' wrote Collins. 'Only time will tell how deep and how far that power will take us. I am willing to bet that the best is yet to come.'

'We wish to suggest a structure for the salt of deoxyribose nucleic acid (DNA). This structure has novel features which are of considerable biological interest.'

Francis Crick and James Watson

27 How to find other universes

- Cosmic coincidences
- The anthropic principle
- Six special numbers
- What might be out there?
- Enter quantum mechanics...

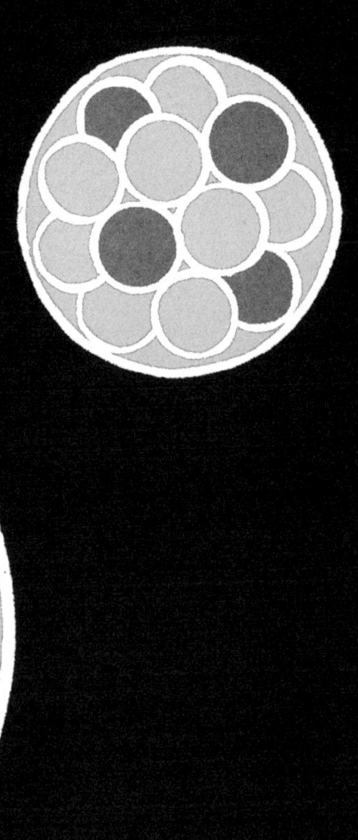

All the complexity we see today depended for its development on a set of very specific physical conditions at the start of the Universe. If any of these conditions had been even slightly different from what we measure today, you would not be here to read this book. Scientists don't like things set up just so without any reason, and have had to go outside our Universe to explain these apparently unlikely coincidences. Far from being special, they argue, our Universe might be just one of many. It may be that every possible universe that can exist does exist.

Cosmic coincidences

Earth might have ended up orbiting the Sun a little closer or a little further away than it does now. The oceans would have boiled away into space in the former case and frozen solid in the latter. Without liquid water, there is no life. Billions of years ago, early in the Solar System's life, the massive gas planet Jupiter did a good job of mopping up the asteroids and other debris that was zipping around. That selfless behaviour saved Earth from a bigger barrage than it was already facing and stopped the early molecules of life that were developing on the planet from being obliterated. Without the Moon, Earth's orbit around Sun would be highly unstable. It would tumble around, causing catastrophic weather patterns that would kill any delicate life forms. No other planets in our neighbourhood have such a large satellite at just the right distance. The Solar System itself is also at the right place: a good distance away from the harsh radiation field surrounding the black hole at the centre of our galaxy but not so far so that it could not have created some of the heavier elements necessary for life.

There are plenty more coincidences at the subatomic scale. Change the relative strengths of the four fundamental forces (gravity, electromagnetism, and strong and weak nuclear interactions) just a tiny amount, and the Universe suddenly looks very different. Atoms don't form or stars burn through their hydrogen fuel too quickly for life to evolve into something intelligent on any nearby planets. The Universe flies apart too quickly or DNA cannot exist. It is impossible to imagine that life exists.

Astronomers have a name for these coincidences and accidents: the Goldilocks zone. Just like the fairy story, the Universe is not too much of one thing, not too much of another, but just right. It all seems too coincidental to be random. How did the blind laws of physics manage to construct all of this on such a monumental scale?

our Universe

parallel universes

identical parallel universe

In the level 1 multiverse, parallel universes are simply regions of space that are too far away for us to have seen yet. On average, our Universe is 10 to the power of 10^{118} metres away from an identical universe.

The anthropic principle

The idea that the cosmos has somehow been fine-tuned to produce life is known as the anthropic principle. In one version of this idea, there is an actor behind the fine-tuning, who set the relative strengths of the forces, organized how matter is made and set off the Big Bang in exactly the right way. This position is known as the 'strong anthropic principle'. The 'weak anthropic principle', on the other hand, states that we would not be around to ask these questions if the Universe was not set up for life. 'The argument can be used to explain why the conditions happen to be just right for the existence of life on the earth at the present time,' wrote physicist Roger Penrose in *The Emperor's New Mind*. 'For if they were not just right, then we should not have found ourselves to be here now, but somewhere else, at some other appropriate time.'

Six special numbers

The anthropic principle highlights the number of things that have to go right for complex structures such as intelligent life to emerge. Cosmologists have put forward the idea that we live in one of many millions of parallel universes. The laws of physics are different in each different universe and most of the universes will not have evolved anything like a life form. British astronomer Martin Rees has identified six numbers, constants of our Universe, which must all have exactly the right values for life to evolve.

The first number, Epsilon, has a value of 0.007. It is an indicator of the strength of the strong nuclear force that keeps atomic nuclei together. If Epsilon were instead 0.006, it is unlikely hydrogen atoms would ever have fused to make helium, in stars or elsewhere. If it were 0.008, all of the hydrogen created at the start of the Universe would have quickly turned to helium and there would be none left to fuse together in stars.

Omega is a measure of the density of matter in the Universe a second after the Big Bang. A higher density by a fraction of just one in a million billion (10^{15}) would have made the Universe collapse in on itself before stars could establish themselves. If the density were the same fraction smaller, particles would have drifted apart and never coalesced to create complex structures.

The electromagnetic force is 10^{36} times stronger than gravity, a number known as *N*. If gravity were any weaker, stars would not have been able to condense from gigantic clouds of hydrogen gas. They would never have been able to produce the immense temperatures and pressures required at their

'The latest theories of quantum gravity count some 10,500 realizations of the Universe, in which the various fundamental constants of nature differ. In this Multiverse, all universes are equally real, although we can only hope to explore our own one. Given the staggering array of alternatives, it is exceedingly unlikely that our observed Universe should even exist.'

Joseph Silk

cores to fuse protons (which repel each other due to their mutual positive charges). No fusion means that stars would not shine and planets would have no source of energy. If gravity were stronger, stars would burn through their hydrogen fuel far too quickly for intelligent life to evolve on any nearby planets. Stronger gravity would also have brought stars, planets and asteroids closer together, increasing the chances of catastrophic collisions.

Next on Rees's number list is Lambda, the cosmological constant, which determines how fast the Universe is expanding. A negative Lambda would contract the Universe into a big crunch. A Lambda several times higher than the assumed value for our Universe means that our cosmos would keep expanding at an ever-increasing rate, stopping matter particles from coalescing and forming stars and galaxies.

D represents the number of spatial dimensions that are needed to support life. We exist in three spatial dimensions and this has allowed complex structures, such as brains, to develop. Three dimensions allows lots of neurons with overlapping nerve connections between them to form a brain, a structure that is hard to imagine in one or two dimensions. In four dimensions, on the other hand, planets are not stable in gravitational orbits and electrons also become unstable in their orbits around atomic nuclei.

Finally there is *Q*, which explains the irregularities observed in the microwave background radiation that can be detected in all parts of the sky, a remnant of the earliest light from the Big Bang. A smaller *Q* than we can measure today would have left the Universe dark and featureless; a larger *Q* would have brought everything together much sooner in the Universe's evolution, and these massive structures would soon have collapsed into black holes.

The scientific way to explain why Rees's six numbers are exactly right in our Universe is to look for a way to remove any uniqueness from our Universe. Martin Rees believes that the best way to explain why these numbers happen to be right in our Universe is, simply, that there are lots of universes. Just as Earth existing in a Goldilocks zone around the Sun implies the existence of lots of other planets in our galaxy that do not inhabit such perfect locations, the cosmic Goldilocks zone implies the existence of other universes. To paraphrase Rees, if a shop has an extensive stock of clothing, it is no surprise to find a shirt that fits.

Testing this idea is well beyond the ability of physics right now, so you might argue that the multiverse does not seem to take us much further than the anthropic principle. However, the ideas around multiverses makes predictions that we could one day potentially test, as improved telescopes and gravity-wave detectors are brought to bear on searching the heavens.

What might be out there?

The cosmologist Max Tegmark has identified three classes of multiverse. Tegmark's level 1 multiverse comprises bits of own Universe that are too far away for us to have seen them. They would be governed by the same laws of physics as us. Since space is more than likely to be infinite, it is likely that everything that is possible becomes real, no matter how improbable it is. Outside the range of our telescopes are regions of space identical to ours with people and planets exactly the same as ours. More than likely, there is an infinite number of these parallel universes beyond the 42 billion light-year horizon that we can currently observe. Each universe is essentially the same as ours (Rees's six numbers would all be the same) except in the very initial arrangement of the matter. Cosmologists have even calculated the distance between you and an identical copy of you given the possible arrangements of atoms in our Universe: around 10 to the power of 10^{118} metres away.

The level 2 multiverse comprises bubbles of level 1 multiverses separated by inflating empty space.

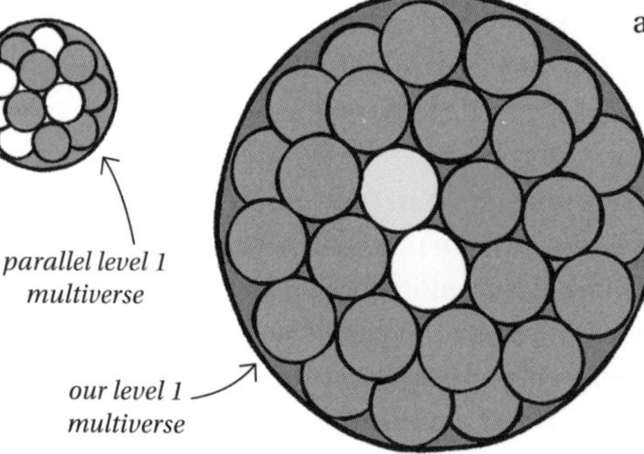

The level 2 multiverse is more in line with what Rees had in mind. It involves tweaking a process that occurred in the moments after the Big Bang. During inflation, the baby Universe grew by a staggering 10^{43} times in less than the blink of an eye. This idea has been established as the reason why the Universe is so uniform, big and flat.

After inflation ended, it was assumed that space continued to expand as normal. But not according to 'chaotic eternal' inflation theory. This idea has it that, though space in general will go on expanding forever more, certain areas will stop stretching and instead form bubbles instead. These bubbles are so far away from Earth that you could never reach them even if you travelled at the speed of light, because the space between our bubble and its neighbours is expanding faster than you could ever travel through it. In different bubble universes, Rees's six numbers have different values. An infinite number of bubble universes allows for Rees's six numbers to be entirely different in each one.

According to Tegmark, when you roll a die, it will land on every possible side in different universes. When the die comes to a rest, it and any observer move into just one of those possible universes, so the observer sees it land on just one side. In this case, the observer finds herself in the universe where the die lands on 1.

Enter quantum mechanics...

Tegmark's level 3 multiverse invokes quantum mechanics to explain how it works. One aspect of quantum mechanics is an intrinsic randomness, the inability for an observer to predict the outcome of an action absolutely. Instead, quantum theory predicts a range of possible scenarios and assigns each of them a probability of happening. In the 'many worlds' hypothesis, each of these possibilities actually corresponds to a specific universe.

Think of a six-sided die. Before it is thrown, there are six possible ways for the die to land and Tegmark would say that each possibility corresponds to a different universe. When the die lands on a particular number, the observer and arena of action move into one of those six universes. In the other five possible universes, which all do exist somewhere, the die landed on one of the other numbers.

The level 3 multiverse can encompass the other two types of multiverse in Tegmark's classification. A universe governed by the 'many worlds' rules splits over time into what look like level 1 parallel universes. Level 2 multiverses can also be accounted for by saying that the quantum split happened in the critical moments after the Big Bang, when the physical constants of the Universe were yet to be fixed. Tegmark's level 3 multiverse leads to some interesting ideas about the nature of time. He writes: 'Most people think of time as a way to describe change ... If parallel universes contain all possible arrangements of matter, then time is simply a way to put those universes into a sequence. The universes themselves are static; change is an illusion, albeit an interesting one.'

28 How to break codes

- Keeping communications safe
- The mathematical age
- The magic of prime numbers: part 1
- The Rivest–Shamir–Adleman algorithm
- The magic of prime numbers: part 2
- Quantum encryption

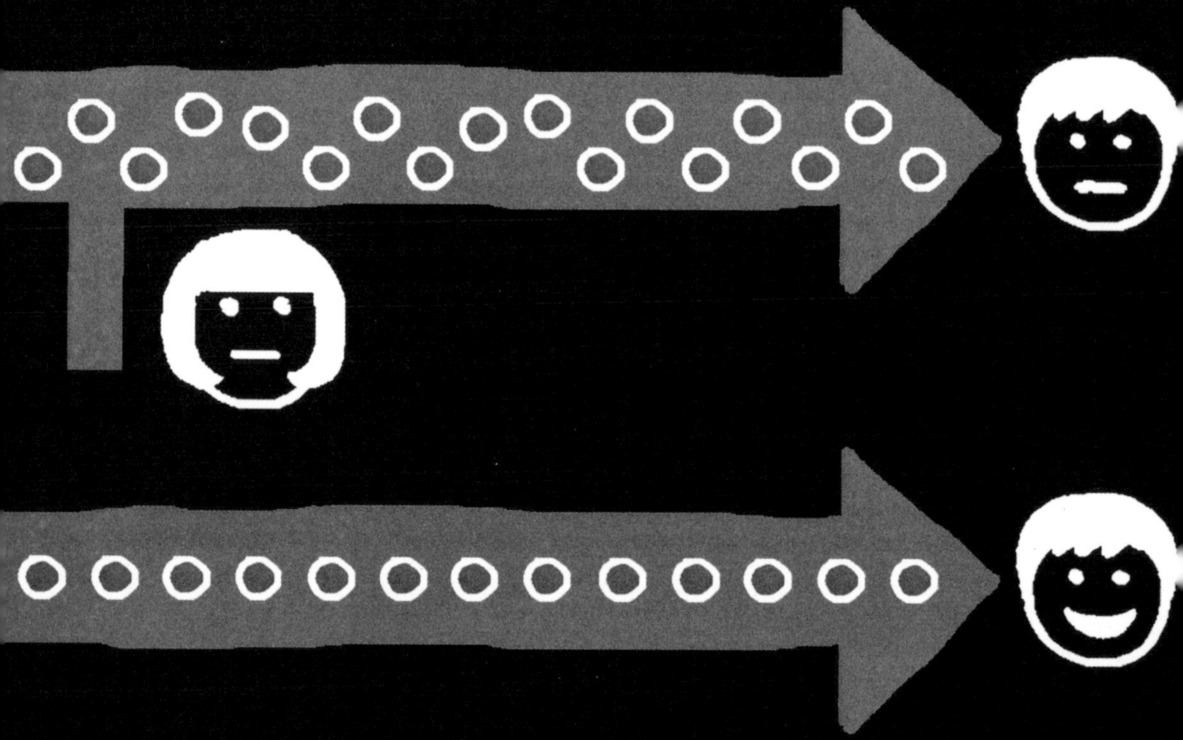

In just a few short decades, the Internet has become an all-pervasive presence in our lives. One of the reasons this has happened is, inevitably, the ever-decreasing cost of the hardware. But another factor, arguably just as important as hardware, is security. For thousands of years, encryption has been used to scramble messages between spies, politicians or military leaders. Today, you use encryption every time you view a web site, log into the computer network at the office or make an electronic payment.

Keeping communications safe

For as long as humans have been able to communicate, they have needed a way to hide the information they were trying to convey. The earliest way of keeping secrets was to write something down – given that very few people could read, it was a simple but effective solution. The Spartans improved on this idea by using a cylindrical device called a scytale. Whoever wanted to send a message would wrap a strip of paper in a spiral around a cylinder and write their words along the length of the cylinder. When the paper was unwound, the arrangement of letters would be meaningless. At the receiver's end, the paper was wound onto a scytale of the same dimensions as the sender's device and the message was revealed. Julius Caesar also used encryption to send secret messages to his allies. He used a formula (or cipher) to replace each letter of the message he was writing with another that was three letters along the alphabet (this is the 'key' that customizes his use of the cipher). So an 'a' would be replaced by a 'd', a 'b' replaced by an 'e' and so on. In that way, the message 'attack Gaul' would read 'dwwdfnjdxo'.

Both the Spartan and Caesarean codes are, clearly, very easy to break. With the former, all you need is a replica scytale. Even if an enemy general did not have specifications for the original scytale, it would not take much time to try out a brute force approach whereby he uses cylinders of different sizes until hitting on the right one. Caesar's code is vulnerable to an attack called frequency analysis. Because every letter in the ciphertext (the scrambled message) maps onto a single letter in the plaintext (the original message), you can use the fact that different letters are used different amounts in a language to predict which letter is which in the message.

A code-breaker will first read lots of plaintext by the message-sender (though compiling texts from other sources such as newspapers or books in the same language would also do) and count each instance of every letter. Then he would do the same counting analysis for the ciphertext and match up the

A B C D E F G H I J K L M

N O P Q R S T U V W X Y Z

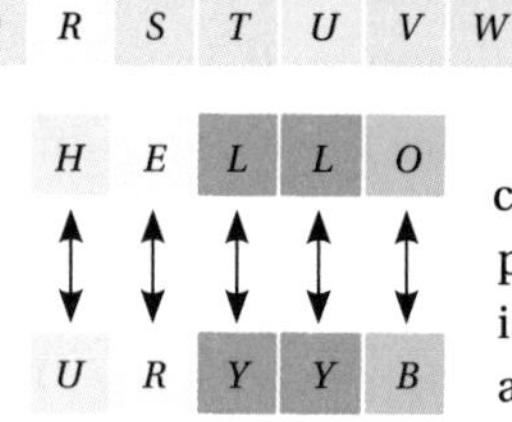

This simple substitution cypher maps the first 13 letters of the alphabet onto the last 13 letters.

distributions of encrypted letters to the reference distribution for the language. If a few words end up wrong, the code-breaker can use knowledge of a language's conventions – if the plaintext is in English, for example, it is likely that the letter 'e' is most common, 'n' is likely to appear between an 'a' and a 'd', and a 'u' usually follows a 'q'. The longer the ciphertext, the more likely that this frequency analysis will work but even messages containing just a few dozen words can be cracked in this way.

Caesar's substitution cipher was the inspiration for more sophisticated encryption mechanisms in the centuries that followed. Perhaps the most famous was the Enigma machine used by the German army during the Second World War, which used rotors and electronics to change the substitution key after every letter was typed out in a message. The ciphers were eventually cracked by a team of code-breakers at Bletchley Park, led by Alan Turing, the forefather of modern computing.

The mathematical age

The modern era of cryptography began in the 1970s, when computer scientists entered the game. Until then, ciphers had been an exercise in language and relied on the skills of whoever was making or breaking the encryption; computers turned the practice into an exercise in mathematics.

One of the problems with pre-mathematical encryption was that both the sender and receiver of the message had to have the key. If you imagine the encryption and decryption of a message as like locking and unlocking a door, the same key would lock and unlock the door. At some point in the past, the sender and receiver would have had to share information about the key so that they could communicate properly. This system works but is cumbersome – the keys have to be distributed everywhere that someone needs to decode messages and the more places the key is sent, the more chances that it could intercepted.

That problem was solved (with a lot help from mathematics) with the invention of 'public-key cryptography'. In effect, this is where one key locks the door but a separate key will unlock it. You can distribute the locking key freely because possessing it does not compromise your security. Anyone who wants to send you a secure message can use your locking key to encrypt their message. But only you can decode it as only you have the unlocking key. In fact, once encrypted, even the sender cannot then decrypt the ciphertext.

The public-key cipher is probably encrypting all of your emails right now. It almost certainly secures your bank transactions online: it is what is happening whenever you see 'https://' in the address bar in your browser.

The magic of prime numbers: part 1

To understand how the public-key cipher works, we need to make a small diversion into some mathematics developed by German mathematician Johann Carl Friedrich Gauss – the 'clock calculator'. Think of a clock face. If the time is 10 o'clock (don't worry about morning or afternoon) and I ask you what the clock will say in five hours, you know the answer is 3 o'clock. In clock arithmetic, this is how addition works – on a 12-hour clock, $10 + 5 = 3$.

You don't need to restrict yourself to 12 hours for your clock. The famous French mathematician, Pierre de Fermat noticed that if you used a prime number, say 'z', of 'hours' for your clock, something interesting happened when you did your calculations. He noticed that, if you take a number on your clock calculator and raise it to the power of z, you always got the number you started with originally. For example, imagine a clock where $z = 7$, in other words there are 7 'hours'. If you start with the number 2 and raise it to the power of seven, you get 128 in conventional maths. But using clock arithmetic, 128 is 2 (that is, you go 18 times around the clock face and get a remainder of 2).

In clock arithmetic using a 12-hour clock, $9 + 4 = 1$

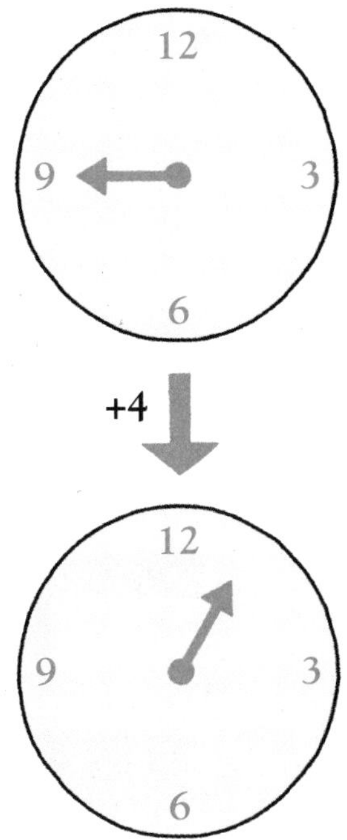

Something about prime numbers meant that the clock started to repeat itself after a certain number of steps. A hundred years after Fermat wrote down his 'little theorem', the mathematician Leonhard Euler worked out that a general clock with M hours, where M is the product of two prime numbers 'r' and 't', would also start repeating itself after a number of steps determined by a set formula of r and t. More than 300 years later, this mathematical curiosity would come in very handy indeed for electronic encryption.

The Rivest–Shamir–Adleman algorithm

The theoretical underpinning for the first public-key cipher was developed in 1976 by Whit Duffie and Martin Hellman of Stanford University in California. A year later, Ronald L. Rivest, Adi Shamir and Leonard M. Adleman, all then at the Massachusetts Institute of Technology, built the first real-world system and called it the 'RSA algorithm' (named after each of their last initials). When you buy something on the Internet, your credit card details are encrypted using a public key on the shop's website, which will consist of two numbers, let's call them M and N. The first number is the product of two prime numbers, r and t, and represents the number of 'hours' on their public clock calculator. The second number, *N*, is called the encoding number. To use the algorithm, you take your credit card number, raise it to the power of *N* on the public clock calculator, and you have your encrypted number, *E*. To decode the message, the web site owner uses his private unlocking key, a number called *D*. This is calculated to be such that, when you multiply the encrypted number *E* by itself *D* times on the clock calculator with *N* hours, you get the original credit card number back.

The magic of prime numbers: part 2

The secret of RSA's security lies in the prime numbers *r* and *t*. If these numbers are relatively small, it is not much of a stretch to work backwards from the steps above to calculate the initial message. But secure websites do not make life easy – the prime numbers they use to build their encryptions are typically thousands or even millions of digits long. This means *M* can typically be billions of digits long and factorizing it into its prime-number constituents, *r* and *t*, is a Herculean, near-impossible task. Even with the fastest computers, it is possible to encode something so that, by brute force alone, it would take more time than the age of the Universe to factorize.

Transmitting messages using quantum cryptography allows the sender and receiver to see if the message carrying the key has been intercepted. If it has, the sender simply resends a different key.

RSA has led to the hunt for ever-larger prime numbers to make Internet systems even more secure. The largest prime number discovered so far, announced in 2008, is $2^{43112609}-1$ and comes in at 12,978,189 digits long. Prime numbers, which are only divisible by themselves and 1, are a mathematical oddity. They appear seemingly at random along the number line. Finding small ones (2, 3, 5, 7 etc.) is obviously easy – just divide each candidate number by all the smaller numbers and see if any of them go in a whole number of times. As the numbers get bigger, however, you need serious computing power to do this: nowadays they are found by massive international projects that network together thousands of PCs around the world. It takes around 25,000 years of computer time in total to find each huge number.

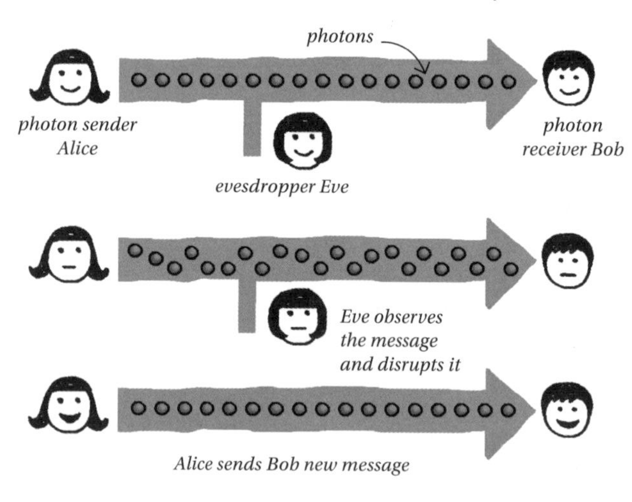

RSA relies on the fact that we can use ever-larger prime numbers to outpace the growing power and speed of computers, so a brute force attack on the encryption system will just take too long to work. But, as history has shown so

'Quantum cryptography makes use of the unusual properties of quantum mechanics to protect encoded information. In trying to listen in on a message sent through a properly designed quantum communication channel, an eavesdropper will inevitably disturb the signal and thereby reveal his or her presence.'

MARK HILLERY

many times, underestimating the future power of computers is a fool's game. At some point, faster computers with yet to be developed mathematical shortcuts could be fast enough to radically bring down the time it takes to factorize prime numbers, thus endangering the security of our entire online world. In effect, the whole thing is an arms race.

Quantum encryption

Fortunately, there might be even more secure alternatives to RSA in the future. This is a way of making encryption virtually foolproof, which includes a way to work out if any rogue agents have been tampering with the message en route to you from the sender. And, because it relies one of the fundamental concepts in physics, it cannot be circumvented.

Quantum cryptography uses the properties of the quantum world to send information securely. In a standard optical fibre communications system, each 'bit' of information (a 0 or a 1) is carried along the fibre by millions of photons, particles of light. An eavesdropper can split off some photons (just a hundred or so) and work out the information they were carrying.

In quantum cryptography, each bit is carried by a single photon. If you remove that photon, it is gone for good and the receiver will never get it. Quantum physics also tells us that you cannot copy a single photon faithfully. If somebody tries, they will introduce changes to the photon, which can be detected by both the sender and the receiver. Spotting those changes can tell you if your message has been compromised.

So far, quantum cryptography has only been demonstrated in small-scale laboratory settings and there are many hurdles to overcome in making it practical, not least in reliably controlling the quantum bits of information. Within a few decades, though, it might become ubiquitous. Perhaps then, insecurity will become a thing of the past.

29 How to live with uncertainty

- The atomization of energy
- The uncertain revolution
- Uncertainty can set you free
- Spooky action
- That's all very well, but what's the point?

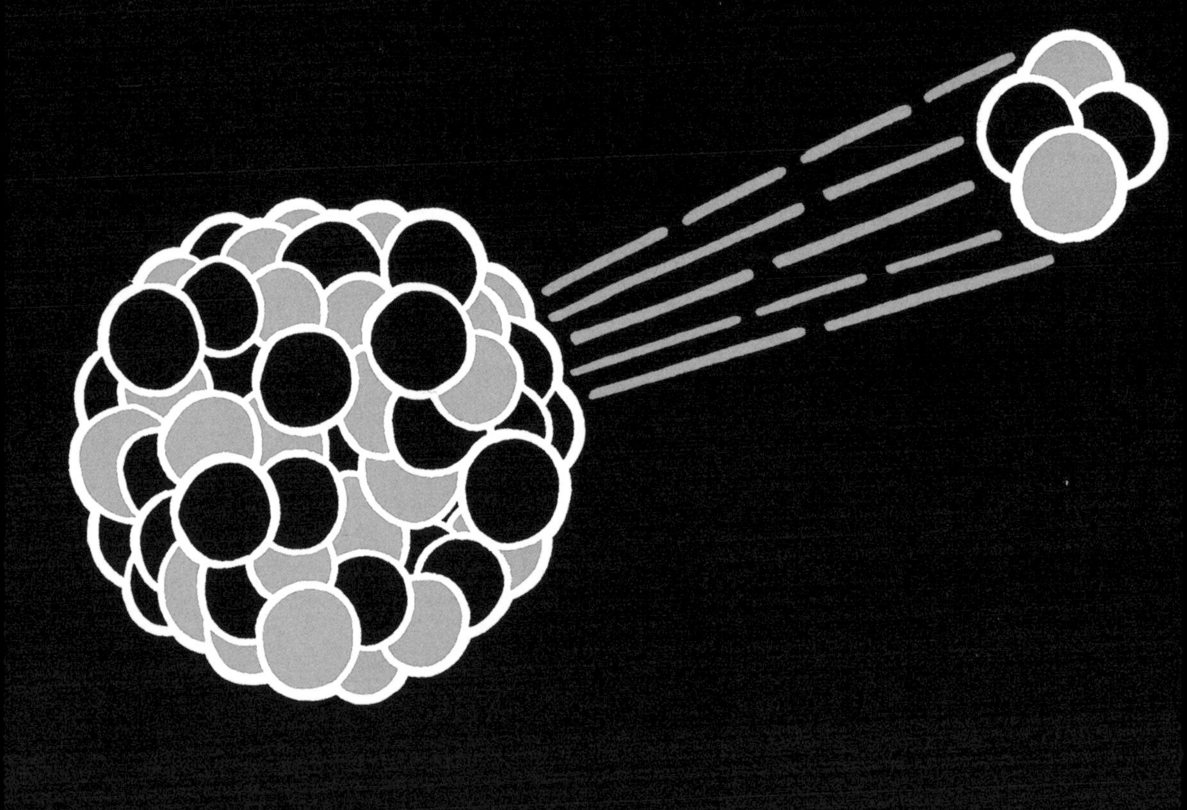

In Newton's clockwork universe, deterministic rules were king: given the starting conditions and what forces were acting on a system, you could calculate its future. This deterministic method was extended to light in the 19th century as James Clerk Maxwell completed his masterful equations showing that light was an electromagnetic wave. There were small details here and there that required some attention but, for the most part, it appeared that physics was almost done and dusted.

The atomization of energy

Less than a year into the 20th century, cracks to started to appear in this deterministic model. Max Planck and Albert Einstein began what would later become the quantum revolution with the suggestion that light might be a stream of discrete packets, or 'quanta', of energy, rather than the continuous wave that Maxwell had proposed. Planck's idea was that the energy contained in an individual quantum of light, known as a photon, was proportional to its measured frequency. Because each photon contained such a small amount of energy, there would be billions in any real-world situation so the light coming from a bulb would look wave-like.

Einstein used Planck's idea to explain a phenomenon known as the photoelectric effect, which occurs whenever light shines onto a metal surface in a vacuum. In 1899, Philip Lenard had found that shining a single colour of light knocked electrons out of the metal atoms and that each electron had exactly the same amount of energy. No matter how bright the light, the energy of each electron was the same – brighter light just made more electrons jump out. When Lenard used light of a higher frequency, though, he found that the electrons leaving the metal also had correspondingly higher energies. Einstein tied Planck and Lenard's work together. He showed that an electron was only ejected from the metal if it was struck by a single photon of the correct energy. If a photon was not energetic enough, it would fail to dislodge anything. Niels Bohr used this result to work out that the electrons in an atom only existed in discrete orbits around the nucleus, each orbit corresponding to a specific level of energy. An electron could move from one orbit to another but only if it could gain or lose a specific amount of energy, in the form of a photon.

From the ideas of Planck, Einstein and Bohr, physicists spent the next few decades developing the foundations of what became quantum mechanics, the most complete description known for the behaviour of the Universe at

'I think I can safely say that nobody understands quantum mechanics.'

RICHARD FEYNMAN

the scale of protons, neutrons and electrons. By the early 1930s, the physics of the very small had well and truly established itself. It revealed a type of science that Newton would never have recognized.

The uncertain revolution

If a light wave can be interpreted as a stream of particles, can particles, such as electrons, also be described in terms of waves? Louis de Broglie wondered this very thing and, in 1924, formulated a way to calculate the 'wavelength' of particles of matter, such as electrons. It took a few years for wave–particle duality to become accepted and it helped that Austrian physicist Erwin Schrödinger had been working on an equation to describe the wave-like nature of the electron, a sort of analogue to Maxwell's equations for light. Schrödinger's wave equation provided one of the first consistent models of the atom, helping to partly explain Bohr's hypothesis that electrons moved around the atomic nucleus in fixed (or quantized) orbits. Rather than pinpointing where an electron is located, Schrödinger's equation describes the probability distribution of where the particle might be at a particular time. This probability wave can propagate through all of space, which leads to the weird but valid result that some of the electrons in the atoms in your finger have a tiny (but non-zero) probability of being on the other side of the galaxy.

By the time de Broglie collected his Nobel prize in 1929, it was clear that all waves could be treated in terms of particles and anything hitherto thought of as particles could also be treated as waves. But wave–particle duality was not the first bit of fuzziness that quantum physics introduced to the carefully crafted deterministic world of 19th-century science.

At around the same time that de Broglie and Schrödinger were blurring the boundaries between waves and particles, Werner Heisenberg was busy inserting uncertainty into the mathematics of quantum mechanics. His 'uncertainty principle' says, in short, that we cannot measure the position and the momentum of a particle in the quantum world with absolute precision. The more accurate our knowledge of one of these properties, the more inaccurate our knowledge will be of the other. Watching a particle or wave in action won't help, since quantum theory says that the very act of observing anything in the quantum world interferes with the system.

To see an object in our everyday macroscopic world, we use our eyes to detect and interpret the light that reflects off it. To 'see' an electron in action, however, is tricky. Bouncing a photon off it will impart some momentum to the electron and perhaps change the path of the particle you're trying to

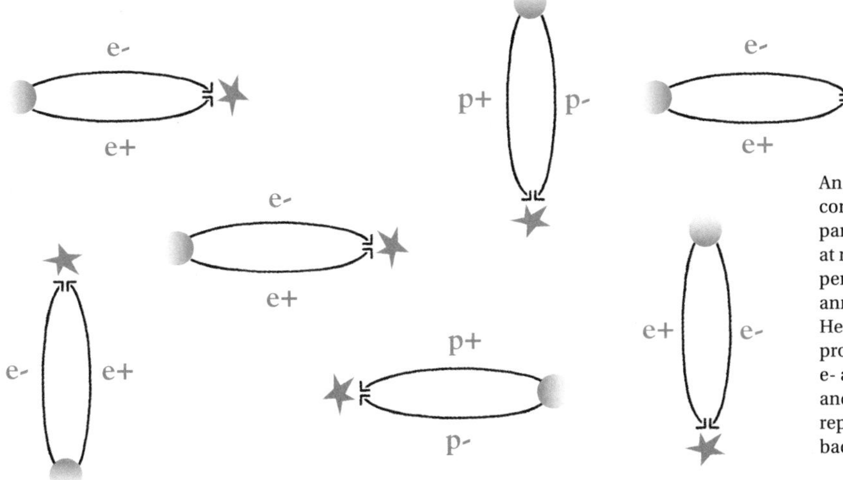

An 'empty' vacuum contains pairs of virtual particles, which appear at random for very short periods of time before annihilating each other. Here, p+ and p- are protons and anti-protons; e- and e+ are electrons and positrons. The stars represent the annihilation back into nothingness.

measure. Either that or, given the high speed at which quantum particles move, the electron won't be in the place it was when the photon originally bounced off it. The very act of observing a system changes that system.

It is worth noting that the wave equation and uncertainty principle are not just mathematical curiosities that can be overcome one day with better instruments or new theories. They reveal the fundamental limits to what we can know about the behaviour of nature. At the quantum level, the best we can do is to work out the probability that something will happen, that two particles will behave in a particular way. Unlike Newton's mechanistic world view, quantum mechanics only tells us one thing for certain: that we cannot be certain of anything.

Uncertainty can set you free

You can look at the uncertainty of quantum mechanics in two ways. If you think it leads to an unsettling loss of logical control, then you're in distinguished company. Einstein may have been one of those who started the quantum revolution, but he didn't approve of the uncertainties, abhorring the idea that nature was essentially random. Indeed, he spent much of the latter part of his life trying to find holes in quantum mechanics. 'I am convinced that He does not throw dice,' he wrote in a letter to his colleague Max Born in 1926. The 'He', in this sense, was Einstein's shorthand for the universal laws of nature rather than any supernatural deity.

Another way to deal with the uncertainty is that it frees nature up from the straitjacket of determinism, and allows things to happen that no rational imagination could ever have thought possible before the 20th century. Quantum mechanics, for example, leads to a curious result when deciding whether a vacuum is really empty.

It turns out that there is an inherent uncertainty in the amount of energy involved in quantum processes and also in the time it takes for those processes to occur. Another way of expressing Heisenberg's uncertainty principle is, in fact, in terms of energy and time – the more accurately you know one variable, the less certain you are of the other. So it is possible that, for very short lengths of time, a quantum system's energy can be very uncertain. If it happens quickly enough, particles can even appear in a vacuum, provided they disappear when their allotted time is up. These 'virtual particles' appear in pairs (an electron and its antimatter pair, the positron, for example) for a short period and them annihilate each other. A vacuum, according to quantum theory, is not empty at all but seething with pairs of virtual particles popping into existence and then vanishing.

Uncertainty also explains a form of radioactivity known as alpha decay. Alpha particles, which are two protons and two neutrons, are usually bound to a much bigger nucleus. To escape they need energy to overcome the bonds tying them in place. The uncertainty principle comes to the rescue. The alpha particle has a very well-defined velocity (as does the nucleus), so that means its position is not so well defined. There is a small, but not zero, chance that the particle will, at some point, find itself outside the nucleus, even though it technically does not have the energy to escape. At this point the alpha particle makes an escape and we observe radioactivity.

Spooky action

Schrödinger's wave equation tells us where a single quantum system (an electron, say) might be in space and how its probability distribution changes over time. In 1935, Schrödinger wondered what would happen if two or more quantum states became 'entangled', in other words you could use the same wave equation to describe them all. Entanglement means that the systems would share the same states. Whatever happens to one of them would also happen to the other. So far so classical – lots of systems can be joined together and, as long as they communicate somehow, they can share the same fate.

According to Heisenberg's uncertainty principle, alpha particles in a nucleus have a small chance of finding themselves outside the nucleus. When they do so, we observe radioactive decay.

But here's where quantum entanglement gets weird. The sharing of fates for these systems works whether the particles are a few metres apart or a few billion light-years apart. Change something about one particle in the system and its entangled pair can change at the same instant.

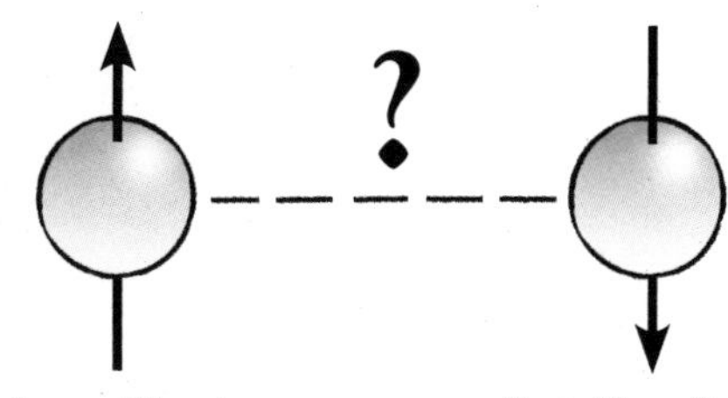

If two particles are entangled, any change affecting one will also affect the other, even if it is on the other side of the Universe.

Something this strange sounds like it should break some law of physics. In a sense it does – one of the unbreakable parts of Einstein's theory of relativity is that nothing can travel faster than light. If two quantum systems are a light-year apart, it would take at least a year for them to communicate using any normal means. Quantum entanglement seems to ignore that rule. It is another aspect of quantum mechanics that Einstein didn't like. When he heard about entanglement, the great scientist called it 'spooky action at a distance' and cited it as yet another reason that quantum physics was too strange to be correct.

Einstein's scepticism was misplaced, however. Experiments in the 1980s observed entanglement in photons for the first time and more recent tests have shown that the 'communication' between entangled quantum systems happens at a speed that is at least 10,000 times the speed of light. Terence Rudolph, a physicist at Imperial College London wrote in the journal *Nature* that, 'any theory that tries to explain quantum entanglement by invoking a transmission mechanism will need to be very spooky – spookier, perhaps, than quantum mechanics itself.'

That's all very well, but what's the point?

As quantum mechanics took off in the early 20th century, some physicists began to worry about what it all meant. Is an electron really a wave? What exactly does Schrödinger's wavefunction represent in the real world, outside its mathematical function? But the success of quantum mechanics, despite uncertainties, has been unparalleled. It has been studied and expanded relentlessly for more than 80 years, yet it is still the frontier of physics research – quantum theory forms the basis for the experiments at the Large Hadron Collider at Cern in Geneva, for example.

Back in the real world, scientists' decisions to work out the strange implications of quantum mechanics rather than get overly fixated on the unknowns has paid off handsomely. In a *Scientific American* article celebrating 100 years of quantum mysteries, Max Tegmark and John Archibald Wheeler wrote, 'Quantum mechanics was instrumental in predicting antimatter, understanding radioactivity (leading to nuclear power), accounting for the behaviour of materials such as semi-conductors, and describing interactions such as those between light and matter (leading to the invention of the laser) and of radio waves and nuclei (leading to magnetic resonance imaging).' So how best to live with all this uncertainty? It turns out that you've been really quite good at living with it all along.

30 How to know yourself

- The Hard Problem
- The scientific study of consciousness
- Just a complex machine?
- The axioms of consciousness
- Conscious robots
- Why make robots conscious?

Where are you? Not in a geographical sense but, rather, where is your 'self'? Do you have free will? Are you sure? Consciousness is the last outpost of pure mystery in our scientific understanding of the brain. It is an area that many researchers have stayed away from until recently, arguing that these concepts are too philosophical for the experimental nature of science. But as any neuroscientist or biologist knows, there is a big hole in our scientific understanding of the way mind and body relate.

The Hard Problem

The 'Hard Problem' of consciousness was formulated by the philosopher David Chalmers. Assuming we can understand everything about how the brain works, Chalmers said we would still have no idea why there was anything like experience generated by this stuff. In other words, why is there consciousness in the Universe at all?

Brains store information in networks of neurons, or brain cells. There are more than 100 billion neurons in the brain. Groups of them fire in a certain sequence when a person sees or experiences an object, and that firing pattern is how a memory of the object is formed. The information stored by networks is not just memory. It could be instructions on how to regulate the level of hormones to control body temperature, or it could be an instinctual response to something that looks like a predator. Interconnections between networks of neurons create feedback loops to produce the property we experience as consciousness, but precisely how this happens is a mystery.

Is this a duck facing left, or a rabbit facing up? We make decisions such as this all the time as we give meaning to the information arriving at our senses.

Scientific tools and techniques are slowly attacking the problem. Take the issue of free will. In 2008, neuroscientists showed that the decision to move a finger is preceded by several seconds of processing in the brain, most of it outside conscious control. John-Dylan Haynes of the Max Planck Institute for Human Cognitive and Brain Sciences in Leipzig, Germany, started by placing volunteers in a functional magnetic resonance imaging scanner. This machine takes pictures of the brain every few seconds to build a two or three-dimensional image of real-time brain activity. The researchers asked the volunteers to press a button with either their right or left index finger whenever they wanted to. At the same time, the volunteers watched a series of letters flickering on the screen in front of them and they were asked to remember the letter that was presented on the screen at the moment they decided to press the button. The results, published in *Nature Neuroscience*, showed how late on in the decision-making process our conscious minds get in on the action. A full second before the button was pressed, the decision had been made by the volunteers' conscious minds. But when Haynes looked at the activity in different parts of the brain in the moments leading up to the button being pressed, he noticed tell-tale signatures of activity in specific areas of the brain that are thought to be involved in executive function and self-processing. And these areas were active up to 10 seconds before any decision to press a button entered the volunteers' consciousness. Does that mean free will does not exist?

The scientific study of consciousness

Scientists like to start with a well-defined problem when studying something they don't yet understand. This has held up work in consciousness because the subject itself is so hard to define. But that's changing. 'The clear, unambiguous definition [often] comes towards the end of the process of scientific inquiry,' says Anil Seth, co-director of the Sackler Centre for Consciousness Science at the University of Sussex. It is possible to distinguish between conscious level – the scale between being completely asleep and completely awake – and conscious content, the actual components of a given experience. Seth adds: 'There's another important distinction between primary consciousness – the raw components of an experience at any given time – and what people variously call higher-order or reflexive consciousness. This is the part of our experience that maps onto our concept of "I". There is an experiencing subject for all these experiences we're having. I have a suspicion that the Hard Problem might not be a problem at all once we've really got to grips with the easy problems.'

Our conscious awareness has many different levels. Some, such as wakefulness, are easy to measure. Others, such as persistent vegetative state, are harder to evaluate: a person may be totally unable to communicate yet still maintain a significant level of awareness.

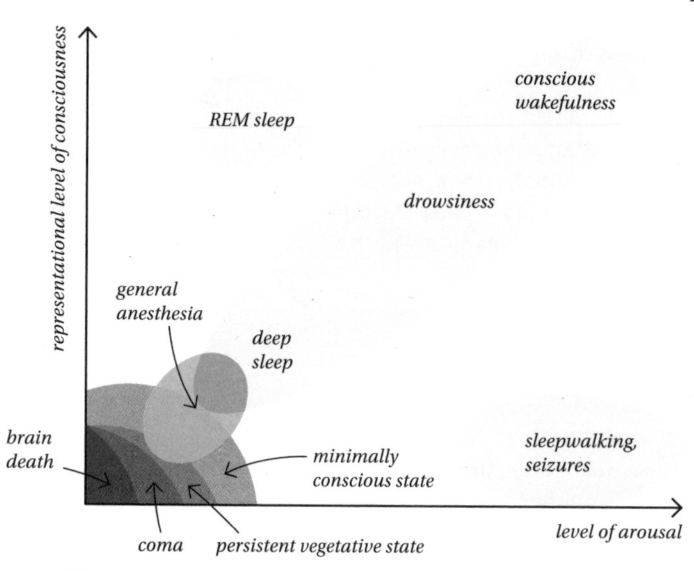

Just a complex machine?

'A brain has up to 60 identified areas,' says Igor Aleksander, an electrical engineer who is trying to build machines that recreate consciousness. 'Within this complex system, we can start discovering what the mechanisms

'Consciousness exists, we know when we're conscious and when we're not. We can start to study those differences, in the same way physicists have made an awful lot of progress without worrying about why there's a Universe at all.'

Anil Seth

that support deliberation are. Consciousness must come out of these interactions.' Aleksander was part of a small group of scientists who began looking at consciousness from a mechanistic viewpoint several decades ago. The work only began to gain wider acceptance and interest, however, when DNA pioneer Francis Crick wrote *The Astonishing Hypothesis* in 1994. Crick concluded that consciousness is simply a product of the interaction of neurons. Studies on the brain in the early 1990s began to suggest that consciousness must be driven by particular mechanisms, since researchers often saw that when people's brains were damaged, their consciousness, their view of the world and their place in it, would often become distorted. 'We started building machines with which one could study hypotheses about the creation of consciousness,' says Aleksander. The study is useful for both engineers and biologists. On one end of the continuum are those who want to build machines that behave in ways that a conscious organism might – computers that could interact with people's unpredictability (on a telephone booking line, for example). Others work at the other end of the research, using models of machine consciousness in a bid to understand how consciousness works in animals.

The axioms of consciousness

Aleksander formalized his mechanistic vision of consciousness into five conditions that an organism (real or artificial) had to display before it could be called conscious. They are a sense of place, imagination, focused attention, forward planning and emotion.

The first axiom comes from the idea that we feel at the centre of a world that is 'out there' and we have the ability to place ourselves in that world. Imagination, the second axiom, is an expression of human ability to 'see' things that we have experienced in the past and conjure up things we have never seen. Focused attention says that we are not just passive reflections of what is happening in the world, that we can focus our attention, and we are conscious only of the things that we attend to. Forward planning allows us to carry out 'what if?' exercises and scenarios of future events. Finally, with emotion, we have a guide for us to recognize what is good or bad for us. In biology, the axioms translate into a way of understanding how brain damage can distort a person's consciousness. 'In axiom one, eye movement is important,' says Aleksander. 'Parkinson's being a disease of movement through lack of dopamine and all that, the eyes don't attend to things as well. That leads to a distorted consciousness.' In a similar vein, Anil Seth is looking at whether psychiatric disorders can be defined as disturbances of conscious experience.

Conscious robots

Fictional robots always have a personality: Marvin was paranoid, C-3PO was fussy and HAL 9000 was murderous. But reality is disappointingly different. Modern robots are dumb automatons, incapable of striking up relationships with their human operators. They are capable of following instructions but have no capability to 'think' for themselves. Several years ago, Aleksander was describing his work on artificial vision to a group of children. When he finished, one unimpressed six-year-old stood up and addressed the engineer matter-of-factly. 'Seeing is the easiest thing in the world,' she said. 'I do it, my little brother does it.' It turns out that the things that humans do most easily, such as recognize faces or interact naturally with other people, are the hardest things to replicate in machines.

Building an artificial consciousness requires some sort of 'brain'. The simplest way to do this is mimicry – the first attempts to do it used electrical circuits soldered together to simulate networks of neurons. Modern neural networks are far more sophisticated and run on software that can 'learn' behaviours based on what they do most. As they become more sophisticated, could they eventually satisfy the axioms of consciousness? Aleksander says that this is a category mistake. 'It's like saying is a horse like a dog? In some ways it is, in some ways it isn't. In the business of consciousness, the most vital question people often forget is that if you've made a conscious object, the question is what's it conscious of?' A conscious robot, for example, should be aware of being a piece of tin with silicon circuits just as a person is conscious of being a biological organism. If an artificial device sophisticated enough to hold a discussion with a person insists that it is conscious like a human then, says Aleksander, it is malfunctioning.

Science fiction is full of intelligent robots and computers that somehow go wrong and end up hurting people or worse. Aleksander says that these are engineering problems rather than ethical dilemmas. 'A properly functioning conscious machine is going to drive your car and it's going to drive it safely.

It will be very pleased when it does that, it's going to be worried if it has an accident. If suddenly it decides, I'm going to kill my passenger and drive into a wall, that's a malfunction,' he says. 'Human beings can malfunction in that way. For human beings, you have the law to legislate, for machines you have engineering procedures.'

Why make robots conscious?

How would consciousness be useful in a machine? It may be advantageous for a robot exploring a distant planet to be conscious of dangers, to be pleased with its own successes. Computer scientist Owen Holland at the University of Essex believes that a conscious robot would have to build up internal models – one for the 'self' and another for the world around it. Humans do this kind of thing without thinking. In order to plan realistically, for example, we need to know not only what our physical limitations are (what we could do) and what is the best choice if we consider our options (what we should do) but also what we are likely to choose (what we would do).Holland uses the analogy of a recovering alcoholic. 'He can go to the bar next door or the tobacconist half a mile away. If he fools himself that he can just go into the bar and just buy the cigarettes, [he knows that] he will buy a drink. In order for him to plan successfully, he needs to know what he would do.' These internal and external models would be the basis from which the robot would build experience. By experimenting with how its own body reacts with the world around it, the robot will learn what is beneficial to it and what is not. Consciousness in a machine would make for better robots. Instead of blindly following their programming, conscious robots would be better able to react dynamically to their environment, adjusting their behaviour on any information they could gather.

In our conscious representation, we constantly have to decide what is foreground and what background. This image may be a vase or two faces, but it cannot be both at the same time.

For true signs of consciousness, Holland will be looking out for some unusual signs in his robots. 'Consciousness is actually very imperfect. The view we present to ourselves and the view of our own past is hugely distorted. If I find the system actually has the same faults as [our] consciousness then I think we might be able to say that its quite likely that human consciousness has its origins in a system of this type.' No-one yet knows what consciousness is or how it emerges but, by trying to replicate it, it is likely that an unlikely coalition of biologists and engineers will be the ones to stumble upon its secrets.

31 How to spot a pseudoscientist

- Who believes what?
- The endarkenment
- Is the problem that we don't know how science works?
- How to be a skeptic
- The limits of knowledge

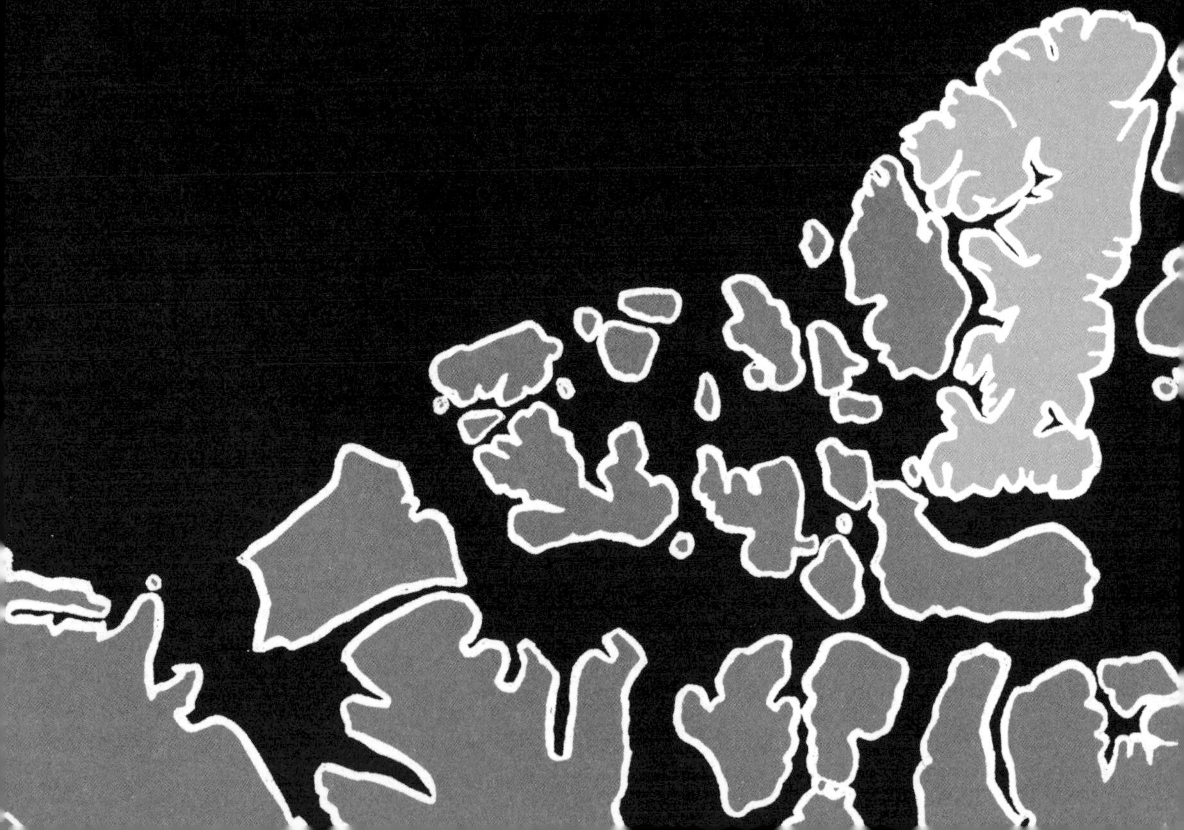

Scientific thought is a beautiful thing. In just a few centuries, this self-correcting method of gathering knowledge has created much of the world around us. It is a simple process: think of an idea, test it with an experiment, draw conclusions and refine your idea. Repeat until your idea is either proved wrong or it adds to our growing understanding of nature. But something sinister has been happening in the past few decades. Despite the progress, despite the untold benefits, belief in pseudoscience is on the rise.

Who believes what?

One of the most developed pseudoscientific ideas around today is intelligent design. This latter-day form of Biblical creationism suggests that life is far too complex to have evolved by chance. Every creature on Earth must, therefore, have been designed and created by some intelligent entity. Despite the continual and evidence-based refutations by evolutionary biologists, intelligent design remains strong, particularly in the United States. One of its central claims is that, if creatures evolved as proper scientists say they did, where are the intermediate forms in the fossil record? Where are the things that look something like a half-way point between two other creatures?

'Science is a way of trying not to fool yourself. The first principle is that you must not fool yourself, and you are the easiest person to fool.'

RICHARD FEYNMAN

If you want to call it such a thing, the *Tiktaalik roseae* is as intermediate as it gets. Found in 2006, the fossil of this crocodile-like animal showed that it had lived in the Devonian era, 417m–354m years ago. It had a skull, neck and ribs similar to early limbed animals known as tetrapods, as well as a more primitive jaw, fins and scales more akin to fish. The animal was a predator with sharp teeth and a body that grew up to 2.75 m (9 ft) long. Like *Archaeopteryx*, the fossil that bridged the gap between reptiles and birds, *Tiktaalik* was an iconic animal that showed evolution in action. Here was an important piece of the puzzle that described one of the most significant rites of passage in our evolutionary history – our emergence from water to land. But rock-solid evidence is never enough to sway the intelligent design lobby. Where there had once been a single fossil gap between two types of species, they will argue, the discovery of *Tiktaalik* now means there are two gaps.

A study carried out by the US National Science Foundation in 2006 found that belief in pseudoscience rose significantly during the 1990s and even into the early part of the 21st century, before falling slightly by 2005. Nevertheless, around 75 percent of Americans claimed to hold at least one pseudoscientific belief. At least a quarter of the population believes in astrology with a third of the population believing it is 'sort of scientific'.

This is the skeleton of *Archaeopteryx*, which lived 147 million years ago. It represents a transitional fossil between dinosaurs and birds, with adaptations for flight such as feathers and larger, more powerful eyes.

In Europe, the figures are more worrying: 53 percent of those polled in 2001 thought astrology was 'rather scientific'. In another Eurobarometer poll in 2005, 41 percent of respondents rated astrology as 4 out of 5 on a scale of how scientific it was (with 5 being most scientific), the same value given to economics. Europeans were also more likely than Americans to agree with the statement: 'some numbers are particularly lucky for some people'.

The endarkenment

The prominent pharmacologist at University College London, David Colquhoun, argues that the past 30 years have been an age of 'endarkenment', a reverse of the centuries enlightenment during which science was born. Writing in semi-despair, he says: 'It has been a period in which truth ceased to matter very much, and dogma and irrationality became once more respectable.' Can things really be so bad? Evolutionary biologist and arch rationalist Richard Dawkins is also pessimistic. 'Science has sent an orbiter to Neptune, eradicated smallpox and created a supercomputer that can do 60 trillion calculations per second,' he said in a programme exposing pseudoscience, *Enemies of Reason*. 'Science frees us from supersition and dogma and allows us to base our knowledge on evidence. Well, most of us.' Today, he says, reason has a battle on its hands. Irrational thinking is a 'multi-million pound industry that impoverishes our culture and throws us new-age gurus who exhort us to run away from reality ... I believe it profoundly undermines civilization.'

In a speech to the Royal Society in 2005, the then president Robert May warned that the core values of science, among them free enquiry, were 'under serious threat' from resurgent fundamentalism:

> *Ahead of us lie dangerous times. There are serious problems that derive from the realities of the external world: climate change, loss of biological diversity, new and re-emerging diseases, and more. Many of these threats*

are not yet immediate, yet their nonlinear character is such that we need to be acting today. And we have no evolutionary experience of acting on behalf of a distant future; we even lack basic understanding of important aspects of our own societies. Sadly, for many, the response is to retreat from complexity and difficulty by embracing the darkness of fundamentalist unreason.

Is the problem that we don't know how science works?

It is probably true that many of us don't really know how science works. But, unless you are a scientist yourself, why should you know about all the arguments, the theories that are worked on but never prove correct, the painstaking nature of many experiments and the daily frustrations involved in the minutiae of cutting-edge research? Never mind understanding the process of peer review, the process of analysis by researchers in a field, used to assess all new discoveries and determine whether or not a research paper makes it into a science journal. A 2004 poll in the UK, commissioned by the Science Media Centre and the journal *Nature*, showed that almost 75 percent of the public did not know what this basic aspect of the scientific research process actually was. When we're not even armed with the knowledge of what science is, how can anyone be expected to tell science from pseudoscience? The route to empowerment lies in education.

How to be a skeptic

The American physicist Robert L. Park has a motto for anyone hunting out fuzzy thinking. 'Extraordinary claims demand extraordinary evidence,' he wrote in his book, *Voodoo Science*, echoing the sentiments of the great astronomer and popularizer of science, Carl Sagan, and philosopher and skeptic, David Hume.

The key is that you don't need to be a scientist to think scientifically. Scientists are professional skeptics, who employ a way of thinking that everyone should employ. This way of thinking can quickly allow you to discern whether something is believable or just babble.

Fish in the Late Devonian Period started to venture out of shallow, oxygen-poor waters to begin to exploit the land.

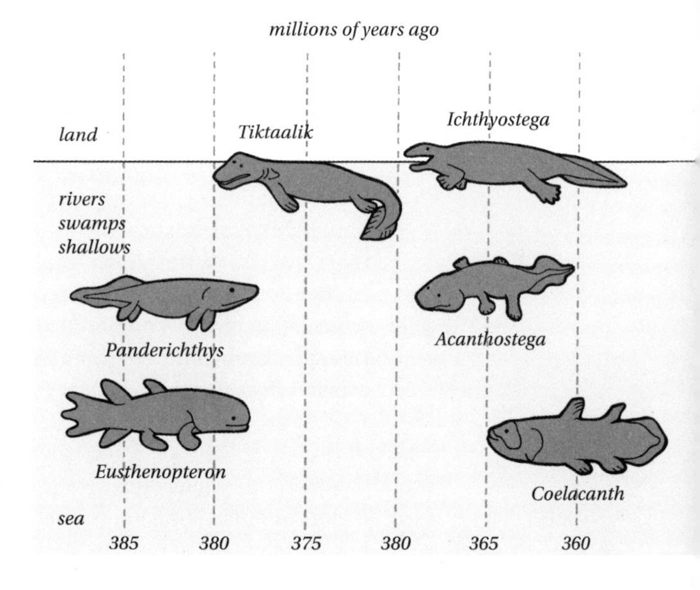

The first thing to look for in pseudoscience is dogma. In its excellent guide to pseudoscience, the UK Skeptics website defines dogma:

> *A dogmatic belief or position is one that is deemed, by its proponents, to be accepted authority; as such, it not to be doubted or disputed. Pseudosciences tend to have evolved very little, or not at all, since the dogma was first established. Any research or experimentation that is carried out in the field is generally done more to justify the belief than to improve knowledge.*

Science is in a continual process of self-improvement. If someone discovers an incontrovertible piece of evidence that flies in the face of accepted theory, scientists have little choice but to throw that theory out and start again. Pseudoscientists work the other way around – they begin with a conclusion about something and then hunt for the reasons why that idea might be correct. This leads to the very bad practice whereby research that contradicts the central idea is dismissed, while supportive evidence is highlighted. And if you want to see the research published in peer-reviewed scientific journals, forget it. Normally pseudoscientific 'discoveries' are pitched directly to newspapers or TV stations, thereby bypassing the quality-control process of normal science.

The quality of evidence is also important: anecdotes are not a universally bad way of gathering information about something, but they are not usually good enough for a proper scientific understanding of a subject. The process by which we generate stories about an event are fallible, and while one set of anecdotes might lead you to one conclusion, another set could very easily lead to an entirely different conclusion. Good science always relies on independently verifiable empirical data.

Another hallmark of the pseudoscientist is confirmation bias. Have you ever thought about someone and then bumped into them on the street or have them call you on the phone? Could it be that that is evidence for a psychic connection between you? Probably not. Rather, this is an example of selective memory – how about all those thousands of times when you thought of someone and there was no phone call and you never saw them even on the same day?

The limits of knowledge

Some pseudosciences challenge the limits of our scientific knowledge. If you cannot explain how something works using current ideas, say some pseudoscientists, it must be the case that we need new scientific theories.

This starts from a grain of truth – no scientist claims science can explain everything yet. But making the case for a completely new law of nature is a huge step. To paraphrase Park, Sagan and Hume's motto, you need some pretty astounding evidence to pull that off successfully. More likely than overturning centuries of physics, though, you've probably made a mistake in your experiment or measured something incorrectly.

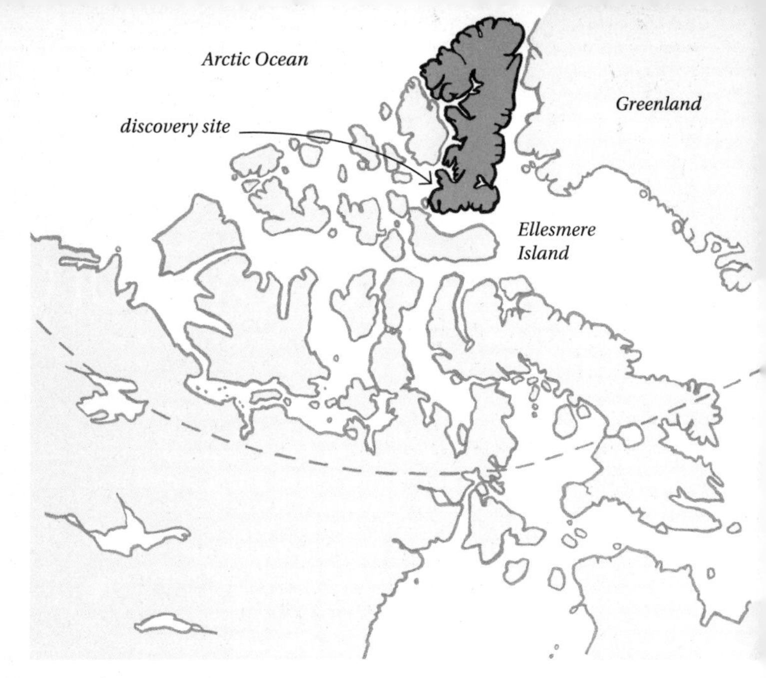

Fossils of the amphibious fish *Titaalik rosae* were found in Ellesmere Island in the Canadian Arctic. During *Titaalik*'s time, this area was part of the continent Laurentia, which was centred on the equator.

Homeopaths think that ailments can be treated by taking an ingredient dissolved in water – cinchona bark for malaria or arsenic for anxiety. Perversely, homeopaths believe that the more dilute a solution is, the more powerful its effects. This means that some of the most 'powerful' homeopathic treatments are diluted to such a degree that there is a vanishingly small probability that even a single molecule of the active ingredient remains in the solution. Homeopaths say it doesn't matter, that the water 'remembers' the ingredients. But they cannot explain the mechanism for their treatments and have no plausible theories.

Pseudoscientists might not want to bother with scientific process or evidence, but they are acutely aware of the authority of science as a way to persuade the rest of us that something is worthy of our time. For that reason, many pseudoscientists erroneously insert scientific words or even entire scientific concepts into their nonsensical claims, safe in the knowledge that the people they want to fool will never check up on them. The most commonly used ideas are those at the unpredictable edges of physics: homeopaths have tried using chaos theory to explain their ideas, while quantum mechanics is often co-opted to explain psychic powers or telekenesis. It's true that we still have much to learn about many scientific ideas. But, as weird as chaos or quantum theory gets, no serious scientist has yet strayed into the paranormal.

And finally, if you ever come across a group that thinks science is a conspiracy, a way of suppressing the truth and blinding you from knowing what is really happening in the world, step away slowly, turn, and run.

32 How to become a cyborg

- Rebuilding the body
- Enhancements for the healthy
- New organs for old
- Neuroenhancement
- Soma for the masses
- Would you do it?

'We can rebuild him, we have the technology,' began each episode of the classic 1970s TV series, The Six Million Dollar Man. *The character, Steve Austin, had implants fitted after he was wounded in an accident. And in real life too, prosthetic limbs and robotic body parts are usually the preserve of people who have lost some function through accident or disease. But if reliable bionic body-parts are available, why not use them to make healthy people even more capable? Imagine soldiers who could run dozens of miles across a desert every day. And how about improving your brain too?*

Rebuilding the body

Seeing through walls or zooming in on faraway objects were among Steve Austin's special talents. For his modern equivalent, however, scientists have been working on ways to restore functional sight to people who have become blind through disease. Degenerative retinal diseases result in the death of the rods and cones, the cells responsible for light detection, at the back of the eye. Worldwide, more than 1.5 million people suffer from a form of inherited blindness called retinitis pigmentosa and, in an ageing population, loss of vision is increasingly common.

Daniel Palanker, a physicist at Stanford University in California, had the idea to bypass the dead rods and cones and to stimulate the cells of the inner retina directly with electrical signals. Previous research had shown this method allowed perception of light, and Palanker built a way to exploit it. His 'bionic eye' is made up of a 3mm chip implanted into the retina and a pair of virtual reality-style goggles containing a video camera. The goggles convert the video pictures into an infrared image. The image is projected onto the retina, where a retinal implant with photosensitive pixels converts it into pulses of electrical current, stimulating the cells in the retina.

In 2008, a limited trial of a similar system was carried out on patients at the Moorfields Eye Hospital in London. Two men in their 50s with retinitis pigmentosa had their sight partially restored using bionic eye implants. They were able to walk around unaided and identify simple objects.

A bionic eye works by transmitting an image from a tiny video recorder in a pair of glasses directly onto the retina, where an implant converts the signal into pulses of electrical current.

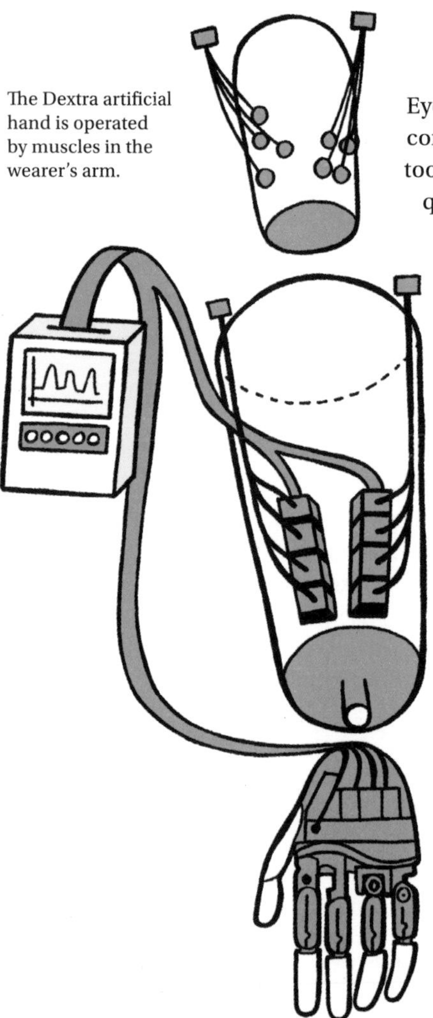
The Dextra artificial hand is operated by muscles in the wearer's arm.

Eyes are not the only body parts that can be helped with computer-aided technology – scientists are improving limbs too. Modern artificial hands, for example, give wearers a better quality of life but they have little of the functionality of the real hands they replace. Current technology might allow a wearer to open and close an appendage or it might give them a hook so that they can manipulate objects, but there is little more functionality than that.

At Rutgers University in New Jersey, William Craelius has invented an artificial hand system, called Dextra. By recording the movement of muscles in the remaining part of the arm as a person thinks about moving their hand, Dextra can control up to three fingers. Different patterns of muscle movement correspond to different movements and, after a few minutes of calibration, the robotic hand is ready for action. Dextra allows the wearer to type slowly or even play the piano. One of Craelius's patients even managed to play the saxophone. Craelius is now working on programming Dextra to carry out more complex, coordinated movements such as grasping a key, opening a door or holding a hammer.

If you ever had trouble with your knee, take heart at the developments by the Icelandic company Ossur. The Power Knee takes all the effort out of climbing stairs or hills thanks to a motor hidden in the joint. Normal prosthetic legs are powered by the wearer swinging and lifting their stump to bend the knee joint, but this can put a strain on the back. The prosthetic leg can also fall out of line with the other leg, making stair-climbing a chore. The Power Knee 'talks' to a sensor worn in the opposite shoe, keeping the legs walking in line.

Enhancements for the healthy

So far, we have talked about enhancements for people who have lost function, but the technologies could, if desired, be adapted for use by people who still have function. If you want enhancements to make humans superhuman, look no further than the military. No cyborg is complete without superhuman strength and Homayoon Kazerooni of the University of California, Berkeley, can help. His technology is a robotic device that a person could wear to enable them to effortlessly carry a heavy load.

'Steve Austin, astronaut. A man barely alive. We can rebuild him. We have the technology. We have the capability to build the world's first bionic man. Steve Austin will be that man. Better than he was before. Better, stronger, faster.'

SIX MILLION DOLLAR MAN OPENING CREDITS

The Berkeley Lower Extremity Exoskeleton (Bleex) fits along the legs and has a frame at the wearer's back to fit a backpack. The maximum load is around 90 kg (200 lb) and Kazerooni claims that the wearer would not feel anything at all. Bleex combines the best aspects of human interaction and decision-making processes with the raw power of machines. It has more than 40 sensors and a hydraulics system that, according to the inventors, work something like the human nervous system. The sensors feed information to a central computer brain, which continually calculates how to distribute the weight so that the wearer feels little or none of it.

The military uses of Bleex are obvious: soldiers who can carry huge loads without getting tired would be more useful on any battlefield. But the inventors argue there are plenty of civilian applications too: firemen who need to climb stairs with heavy equipment or disaster relief workers who need to take supplies into areas where vehicles cannot go.

New organs for old

What happens when a supersoldier wearing a Bleex system gets injured? What if that injury is so bad that it permanently damages an organ? The solution, of course, is a transplant, but these are not simple to come by. The waiting list in any country is long and finding matches can be tricky. In addition, the patient will invariably need to take drugs for the rest of their life to prevent the new organ from being attacked by their immune system.

What if you could grow your own organs, however? Researchers hope to grow new tissue from stem cells to replace damaged material. The logical conclusion of this work would be to grow genetically matched organs, which would not be rejected by their immune systems. So far, scientists have grown tendons, cartilage and bladders from stem cells, but none of these has the complexity of organs, which are structures with dozens of different types of cells. In 2007, a research team at Imperial College London grew part of a human heart from stem cells. They produced a flat sheet of cells that behaved in the same way as a heart valve, a major achievement.

A full organ requires something more: a scaffold on which to grow. Fortunately, the progress here has also been swift, with tiny protein scaffolds developed that can be loaded with stem cells and transplanted directly into a patient. The scaffold, developed in mice by Shuguang Zhang at the Massachusetts Institute of Technology, encourages the stem cells to grow into a three-dimensional structure and is itself harmless to the patient.

New organs created using this method might be a decade or so away, but in the meanwhile, what about an unending supply of animal organs? Anthony Warrens of Imperial College London has been working on 'humanized' versions of pig organs that have been modified to be directly implantable into patients. Pigs are of similar size to humans and can be bred in large numbers. They are genetically similar enough for their organs to work in people, but sufficiently distant to reduce the risk of transfer of infections.

Much of the research into xenotransplantation, the process of using animal organs in humans, was pioneered in the UK in the 1990s, but the work faced problems. The principal one is the body's immune system. Antibodies in the blood will bind to the surface of any foreign organ, such as a kidney, and reject it within hours. The target of these antibodies is a carbohydrate structure on the organs called 'gal epitope'. By creating pigs without this substance in their cells, the chances of rejection are much diminished.

Neuroenhancement

What about the brain? Until the latter years of the 20th century, scientists had scant understanding of how to manipulate or read the brain at the level of detail required to reliably enhance ability. That started to change in 2006, however, when John Donoghue of the brain science programme at Brown University, Rhode Island, implanted a 4mm^2 electronic chip into the brain of Matthew Nagle, a 25-year-old Massachusetts man who had been paralysed from the neck down for the previous five years after a stab wound severed his spine. The implant, called BrainGate, allowed Nagle to control a cursor on the screen and open and close the hand on a prosthetic limb just by thinking about the relevant actions.

'Ever since his pig heart transplant, he loves to do this.'

The implant contains 100 electrodes, each one thinner than a single human hair. It was placed into the area of Nagle's brain that controls motor functions, penetrating around 1 mm into the surface. The device picked the electrical activity between neurons and fed the buzz into a computer, which was able to interpret the signals as meaningful instructions.

Brain implants such as this could one day be developed so that limb muscles can be controlled by thought alone, allowing anyone with spinal cord injuries to bypass the damage. Further ahead, implants could help anyone control electronic objects by thought alone.

Soma for the masses

If you don't fancy surgery to improve your brain's abilities, then why not pop a pill? The idea that an array of easily available drugs could be used to improve memory or increase intelligence is the stuff of science fiction dystopia. In *Brave New World*, Aldous Huxley imagined a whole planet under the spell of a pleasure drug called Soma, which was used by the government to keep people compliant, but could brain drugs also be used for good.

Drugs that work on the brain are already common – many people can hardly begin their days without the mind-sharpening effects of caffeine or nicotine. But brain-enhancing pharmaceuticals that have been developed to treat diseases such as Alzheimer's are likely to find increased use among healthy people looking to improve their perception, memory, planning or judgment. Ritalin, prescribed to children with attention deficit hyperactivity disorder, is often used by healthy people to enhance their mental performance. Modafinil, a drug developed to treat narcolepsy, has been shown to reduce impulsiveness and help people focus on problems as it can improve working memory and planning. Modafinil has already been used by the US military to keep soldiers alert and some scientists are considering its usefulness in helping shift workers deal with erratic working hours.

Would you do it?

All styles of human enhancement raise a stack of ethical considerations, from safety questions to ideas about the kinds of society you might want to live in. Steve Austin's fictional life is far more straightforward than the real questions around the future technologies that will become available to us, to turn us into cyborgs if we wish. You may object most of all to using mind-altering drugs to help you work harder or longer. But, in a world where all your colleagues are doing it, can you afford to be left behind?

33 How to read minds

- Brain scanning revolution
- Reading the brain
- The inner workings of the mind
- Social neuroscience
- Truth detection
- Ethical concerns

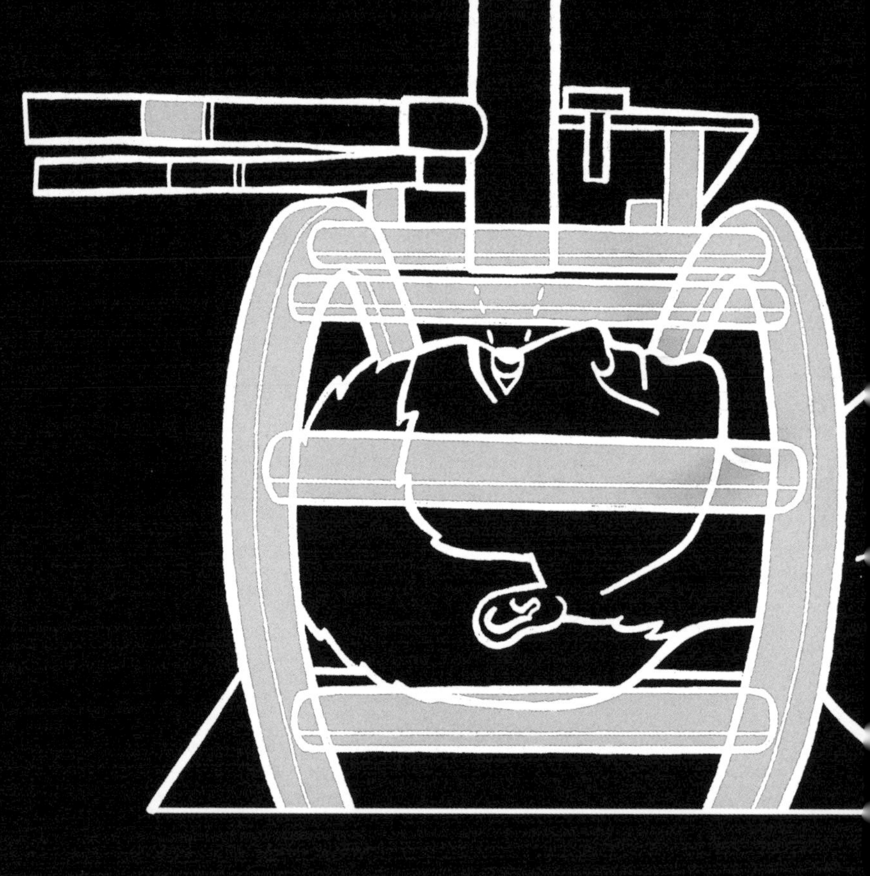

In 2003, neuroscientists carried out a 21st-century version of the Pepsi Challenge. It was found that the volunteers were making decisions based on their memories of a particular drink, as well as taste. By scanning their brains, scientists could build a picture of the factors they considered when declaring their preference. But what are the ethical consequences of such advances? The inner world of our minds may soon no longer be private, as scans reveal our desires, prejudices and even the content of our thoughts.

Brain scanning revolution

Neuroscientists were once restricted to studying microscope slides filled with dead tissues or hunting through scans of brain lesions from people with brain damage or neurodegenerative diseases in order to work out what part of the brain was responsible for which task. In the past, brain science was largely about physiology and structure. Before scanners could watch the brain in action, the deepest questions around how the 'wetware' of white and grey matter in the brain networked together to give us our sensory experience of the world were out of reach.

Successive generations of brain scanners have changed all that. They have been used to assess everything from cellular damage caused by strokes to locating and imaging tumours in readiness for surgery. And, increasingly, they are used to link regions of the brain to their active role in tasks ranging from language to memory, love to fear, morals to our sense of justice.

Reading the brain

Brain imaging machines can be broken down into two basic types: structural, for building up images of what the brain looks like and what is where; and functional, for working out what the bits do. Computed tomography (CT) scans use a series of X-rays from different angles to build up a picture of the head and are often the first port of call in examining any damage to the brain after an acute injury. The electrical hum from the brain's neurons can be measured by attaching electrodes to the skull, allowing doctors to watch changing brain activity as an oscillating graph, called an electroencephalogram (EEG), on a screen. Positron emission tomography (PET) is a technique that uses sensors to detect a radioactive tracer that has been injected into the bloodstream of a patient. The tracer eventually makes its way to the brain and accumulates in particular areas, depending on the chemicals used. For a long time, PET was the best way of identifying the parts of the brain involved in processing a specific task.

For some uses, such as tracking the function of specific receptors in the brain, PET is still the best tool to monitor function. But modern real-time monitoring of the brain is largely carried out by functional magnetic resonance imaging (fMRI), which uses magnetic fields and radio waves to build up a 3-dimensional picture of blood flow in the brain. When brain cells are pressed into action they use more fuel, in the form of glucose, and the blood flowing to them shoots up. By taking pictures of this changing blood flow every few seconds, fMRI scanners can map how the brain is responding to a stimulus in almost real time. Lots of bits of the brain react fairly predictably when carrying out certain tasks and fMRI has allowed scientists to catalogue these. We know, for example, that activity in the brain's insula is associated with disgust and activity in the amygdala, one of the most primitive structures in the centre of the brain, is linked to anxiety or fear.

There is an important caveat: fMRI can tell scientists what parts of the brain are active, but assigning meaning to the findings still requires interpretation, which can differ depending on the scientist and the process used. Say you are lying in a scanner and looking at a picture of a family member. The scanner will show activity in the visual cortex and in networks of neurons that encode a particular memory and emotions associated with the picture. It might also be able to identify activity in the higher parts of the brain that relate to conscious action and thought. The scanner can see you are having thoughts but will not go anywhere in telling anyone what those thoughts are. Scanners can read brains, but not minds.

The inner workings of the mind

Any cursory search of scientific journals will reveal that the past decade has seen a flood in the use of fMRI in hospitals and research labs. In 2008, scientists developed a way to predict what a person was looking at by looking at their brain scans. A computer programme analysed scans taken of volunteers as they looked at a range of black and white photographs. It was able to correctly identify, in nine out of 10 cases, which of the photos the volunteer had been looking at. Random guesswork would have been accurate only eight times in every 1,000 attempts. It may soon be possible to reconstruct an entire picture of a person's visual experience by measuring brain activity in this way.

In an experiment at University College London (UCL), volunteers wearing virtual reality headsets were made to wander around a virtual building. It was found that certain neural networks seem to encode where a person existed in those virtual worlds. Using scans of those cells, the researchers were able to predict where each volunteer was in the building. The research at UCL was part of an investigation into the way that memories are created and stored in a part of the brain called the hippocampus. With a better grasp of that, scientists hope to gain insight into diseases such as Alzheimer's that can destroy memory.

Despite these advances, true mind-reading using this technique is some way off. The person being scanned has to be cooperative, and the computer algorithms that analyse the subsequent images need to be trained with lots of examples of each memory. In the trial, each volunteer was scanned several times in each location in the virtual house.

Brain activity shows up as highlighted areas in fMRI scans, which show how a brain is operating in real time.

Scientists have even read the brains of people unable to communicate in any other way. In 2006, Adrian Owen of the University of Cambridge reported in *Science* how his team had communicated with a 23-year-old woman who had been in a persistent vegetative state (PVS) since being injured in an accident the year before. From a neurological point of view, PVS is a whisker up from a coma – patients show no signs of consciousness and are unresponsive but, unlike people in comas, they have cycles of sleep and wakefulness and some periodically open their eyes. Owen's team spoke to the patient while scanning her brain with fMRI. They asked her to imagine playing tennis on Centre Court at Wimbledon or walking around from room to room in her home. They chose these scenarios because thinking about them employs different parts of the brain in healthy people. Owen's patient with PVS used the same brain regions: the premotor cortex, which governs limb movements, for the tennis task; and the parahippocampal gyrus, which handles maps, for the house task. There was little doubt that, though the woman's body was paralysed with PVS, her brain was capable of responding.

Social neuroscience

As brain maps get bigger and scanners get cheaper, a new discipline is emerging that goes way beyond the hospital or biology lab. It is claimed that the techniques can be used to examine feelings of violence or love, morals, how much trust a person exhibits in others and even their sense of justice. UCL scientists claim to have localized the essence of love to activity in four specific parts of the brain. Joshua Greene at Princeton University looked at moral dilemmas. When asked the classic problem of whether they would push a person in front of a speeding train if that would save the lives of several others, the scientists registered activity in the emotional as well as logical parts of the brain. And another team at the Institute of Cognitive Neuroscience in London found that the amygdala was highly active when people looked at pictures of people they deemed untrustworthy.

A controversial side to the research is the possibility of predicting prejudice. American scientists, led by Jennifer Richeson of Dartmouth College in New Hampshire, claim to have developed a way of detecting racial prejudice. Could such scans become a useful tool for screening potential policemen or public sector workers, for example, to check for any inbuilt prejudices? Not so fast. Scientists do not all agree that potential for prejudice can be so simply defined by brain scans alone. There must also be, some argue, a learnt element to powerful emotions such as racism. In any case, having a predisposition at some deep brain level and acting on it are two separate things.

Truth detection

Brain-reading technology has interesting implications for courts of law if it could be developed into something that might be able to work out whether a person has ever been in a place before, a crime scene for example. CT scans played a role in the 1982 trial of John Hinckley Jr, the attempted assassin of US President Ronald Reagan. PET has been used to diagnose mental illnesses or brain damage that

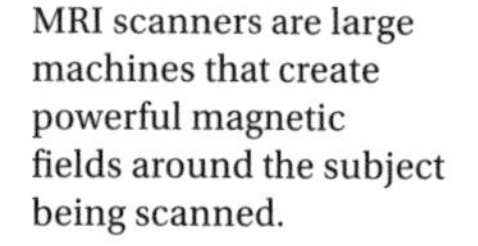

MRI scanners are large machines that create powerful magnetic fields around the subject being scanned.

'Perhaps we've reached the point where this kind of information [brain scan] is distinctive from other kinds of information insurance companies can request such as X-rays.'

Sandy Thomas

might partially explain a defendant's criminal behaviour. fMRI made its debut in a US court in 2009 in the defence of Brian Dugan, who faced a death sentence for murder. Kent Kiehl, of the University of New Mexico in Albuquerque, wanted to see if it was possible to show that Dugan had the brain of a psychopath. Critics argue that, since fMRI is a technique used to compare average brain activity in groups of people, it is tough to interpret the meaning of scans for individuals.

Ethical concerns

Neuroscientists are beginning to grapple with the potential uses and abuses of their work, much in the same way that geneticists have had to grapple with the consequences of their work in discovering predisposition to disease. Adrian Owen's work on PVS raises medical and legal questions. How do you know whether someone really is brain dead and therefore beyond help? What if your patient is aware of their situation at some level, but unable to communicate with anyone around them? Is it still ethical to turn off life-support without the patient's consent? Donald Kennedy, editor of the journal *Science*, believes that brain scans are too personal to be placed in the hands of big business. Because these scans can indicate someone's potential to suffer mental conditions in later life or, as the technology improves, provide clues to their morals or other behaviour, he says no-one other than the owner should know the details.

Even if it is never possible to predict behaviour, there is no doubt that brain scans will be useful in screening for mental illnesses and neurodegenerative diseases such as Alzheimer's, where there are changes in the brain's structure before symptoms appear. Paul Glimcher at the Centre for Neuroscience at New York University argues that legislation that would ban access to people's brain scans must be drawn up sooner rather than later, as insurance companies, in particular, become increasingly interested in them. The ability to diagnose diseases will get better, marketers' ability to sell to us will get better, and what were once our most personal thoughts will one day be accessible to anyone with the right equipment. They may not be able to read thoughts yet, but it will not be long before they can.

34 How to think like an ant

- The most successful species on Earth?
- The world according to ants
- How do you control so many individuals?
- Invisible teamwork
- Insights into the human brain

Ants have perfected a collegiate life. They developed architecture and built farms millions of years before our primate ancestors had even considered walking on two legs. In a typical ant colony, individuals make collective decisions without any central leadership. The study of ants is becoming a major task in biology as we attempt to understand the behaviour of this 'superorganism'. The way ants 'think' is unlike anything else on Earth.

The most successful species on Earth?

Anyone looking for a metaphor for hard work, strength or the power of teams has, at one time or another, used the example of ants. Since the first ants emerged more than 150 million years ago, the insects have made it to every continent except Antarctica, filling every ecological niche as hunters, scavengers or farmers, and evolving into thousands of shapes and sizes. Leptanilline ants are less than 1 mm in length and look like a dusting of pepper; the bulldog ants of Australia can grow to 5 cm (2 in) and each one packs a lethal sting for its victims. Ants eat the same resources as solitary insects but they have been far more successful. Why? 'That's easy,' says Harvard sociobiologist Edward Wilson, 'They live in groups.'

'Wonderful theory, wrong species.'

Edward O. Wilson

Ant colonies range in size from a dozen individuals to millions of insects, mostly consisting of sterile females in specific jobs as workers, soldiers or caretakers, with one or sometimes a few reproductive females presiding over the entire brood. Any males are usually there temporarily, relatively useless drones kept long enough to inseminate the queen, then driven out of the nest or killed quickly afterwards. A queen can store the sperm from a male for over a decade, using it to fertilize millions of eggs. This system has had some astounding results: through evolution, ants have discovered the principles of the Industrial Revolution tens of millions of years before humans.

The world according to ants

One of the most iconic industrial species is the Atta, or leafcutter ant, found in the Americas. A typical colony contains up to 8 million individuals, with the biggest around 200 times heavier than the smallest and each caste specialized for a specific task. The biggest ants cut the leaves from a tree or shrub with powerful jaws that can vibrate thousands of times a second. Another caste takes the fragments back to the nest and a third cuts up the leaf still further. The smallest ants use the bits of leaves to make compost for the fungus that the colony farms for food. They weed the fungus gardens and use antibiotic-producing bacteria to keep the crop free from disease.

The ants' work rate is astonishing. A nest of leafcutter ants can defoliate an entire citrus tree in a day and, in the South American rainforest, the ants typically harvest around a fifth of the annual growth. In a single lifetime, a leaf-cutter colony turns over and aerates 40 tonnes of soil. Ant farms can include livestock. Many colonies keep aphids, tranquillizing them with drugs to keep them docile and 'milking' them with their antennae for a sugary honeydew to use as food. Ants also disperse seeds, particularly in the desert regions of Africa and Australia, and several species of plants are so reliant on the insects that they have evolved special structures on the outsides of their seeds, called elaiosomes, that ants can use as food.

Ponerines are the most diverse families of ants, with more than 1,000 species. They are also the most ancient, and provide an insight into how ants lived before they evolved the highly social societies seen today. Ponerines live in relatively small groups and comprise some of the biggest individual ants. They specialize in hunting just a few types of prey and most of the members of a colony can reproduce, leading to a lot of competition (and therefore low numbers) within a nest.

Trap-jaw ants have the fastest mandibles in the animal kingdom. They can snap shut at speeds of more than 225 km/h (140 mph), killing or maiming prey so that it can be brought back to the nest. The mandibles of larger army ants are used in some parts of the world as surgical stitches: ants are placed along a wound and the insects bite into the sides, pulling the skin together. The bodies are cut away, leaving the mandibles in place along the wound. Some species of army ants use their bodies to plug potholes in the forest floor, creating a flat surface for columns of foraging ants to run back to their nest. Two or more ants band together if there is a particularly big hole. Desert ants navigate outside the nest using visual landmarks and smell. Ants normally excrete chemicals to lay trails to interesting objects or to find their way home, but these volatile chemicals would degrade quickly in arid desert conditions. Instead, desert ants, which roam 100 m (330 ft) or more from their nests, learn the smell of their nest entrance and use it to get home whenever there are too few visual clues.

queen

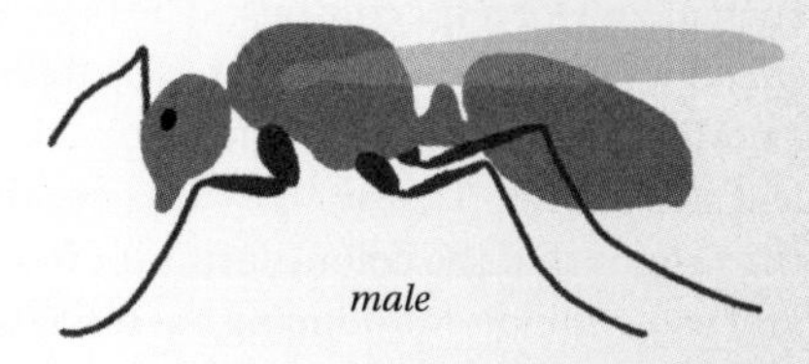

male

All day, every day, the world's ant nests are active: scouting, processing food, fighting and tending to the young. This cohesive working is so important for the wellbeing of the ants that any individuals removed from a colony quickly die. All their work is directed towards the good of the colony.

How do you control so many individuals?

One way ants communicate it is to click and sing to each other, creating sound by rubbing together their body parts. These basic communications can be used against the ants by other species. Some ants will take care of an impostor in their nest, a caterpillar for example, if the impostor makes the right sounds. By playing the sounds of a queen into a colony, scientists have shown that worker ants stood to attention and even greeted the speakers. If the ants, like us, had cognition, an ability to think and reason, they would be able to tell friend from foe and prevent their resources and efforts from being wasted. But if they had cognition they would fail to live in such a social society in the first place. The key to the complexity in the behaviour seen in ant societies is a group of chemicals called pheremones. Around two dozen different chemicals guide the instincts of each insect, telling them which way to turn out of the nest to find food, which of their nest mates is dying and needs to be removed from the colony, which ants need feeding, which are the soldiers, which is the queen and which ones process the colony's rubbish. Rub these chemicals out and the ants get lost.

Pheremones are also at the root of how a colony makes sophisticated group decisions. Take the example of finding a site for a new nest. When a scout discovers a potential site, she will measure attributes such as floor area, light levels and the size of the entrances. If it is a suitable site, she will return to the colony and teach fellow scouts the way to the new site using a process called tandem running, whereby one individual follows the leader by maintaining contact between their antennae. In this way, a number of ants can build up a new nest site and if that number exceeds a threshold, which is called the quorum threshold, the ants will switch from tandem running to picking up their nest mates and carrying them.

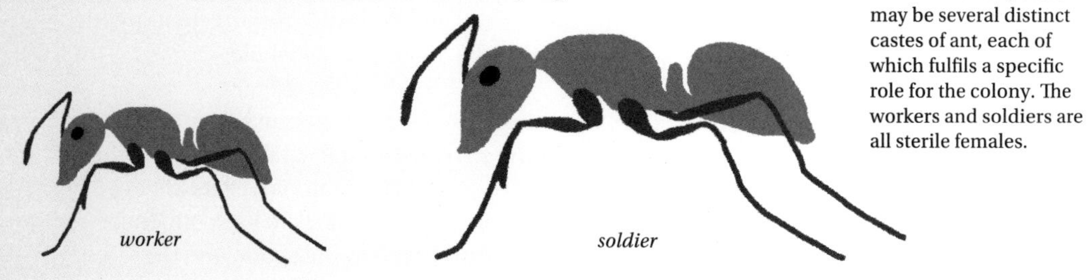

Within one nest, there may be several distinct castes of ant, each of which fulfils a specific role for the colony. The workers and soldiers are all sterile females.

Ant nests can have a complex architecture, with various chambers for egg-rearing, fungus-harvesting and other activities. No single ant has the know-how to build a nest. Rather, the knowledge is held at the level of the entire colony.

Invisible teamwork

Based on environmental cues and pheremones, individual ants make decisions about what to do, whether to forage for food, fight a predator or help an injured nest-mate. But when the number of ants that have made a particular decision reaches a tipping point, the entire colony is committed to the decision. There are different thresholds for different actions: there might be one for leaving the old nest if it is not good, one for accepting a new nest if they begin tandem running.

Ant colonies provide scientists with an invaluable way to gain empirical data around how living in societies has developed. Compare the biological blueprint of an ant society with that of humans and you quickly see that much of human society is built on culture rather than genetics. The basic blueprints for society in our genes are much simpler than those coded within ants.

This sophisticated genetic blueprint inside ants led Edward Wilson, with his long-time collaborator Bert Hoelldobler, a biologist at Arizona State University, to propose an entirely new class of life: the superorganism. 'A superorganism is a closely knit group that divides labour among its members altruistically,' says Wilson. 'There are individuals who reproduce in the group and are promoted to be reproducers, and those that do not reproduce and are workers. This allows the group to function as a giant organism.' Think of a superorganism as a dispersible creature that can stretch out limbs to be in many places at once, going out to forage for food and then withdrawing into the nest after raking up whatever is around. In a book outlining the concept, *Superorganism*, Wilson and Hoelldobler describe their idea by comparing each ant in a colony with a cell in, say, the human body, each one specialized for a task and working (to its own probable death) for the good of the organism as a whole.

Superorganisms can out-compete individuals because they can summon a group that can oust anything trying to take over the food source. A solitary insect might be cautious about risking its life in defending a piece of food it has found and is likely to back off. That is not true of ants. They are willing to fight to the death, and the kamikaze antics do not affect the colony much at all. The group might lose a little strength when workers die in battles but the queen can just produce replacements.

Insights into the human brain

What makes the decision-making powers of ants far more than just a scientific curiosity is that the collective behaviour from what are, at root, chemical-sensing automatons, hints at how our brains might work. Brain cells are individually relatively dumb but, with billions of them working together in our brains reacting to levels of neurotransmitter chemicals, something creative and remarkable emerges. Biologists find similarities when they model ant colonies and collections of neurons. Perhaps the act of making a decision in the human brain (moving your eyes from left to right in response to a threat, for example) is analogous to a threshold number of ants tipping a colony to go into a particular direction. The first bits of evidence to show that this is indeed the way that brains work are already being published in scientific journals. In 2009, computer scientists created models to show that groups of brain cells in the primate brain seem to make decisions in roughly the same way as an ant colony.

The superorganism idea takes the brain analogy further, says Hoelldobler. The action of the ant colony means that it has a sort of collective intelligence. It is a problem-solving unit. Leaf-cutter ants build nests that reach 8 m (26 ft) underground, and have an area of 50 m^2 (more than 500 sq ft). No single ant could build that, or even conceive of such an architecture. But the interaction and behaviour of millions of individuals reacting to particular stimuli that are created by other workers leads to these fantastic structures.

'If they learned to work like us, they'd be dangerous.'

In a way, an ant colony is a problem-solving instrument, with the ants as the individual cells that build into something more than the sum of its parts. And it just so happens that your brain 'thinks' just like an ant colony.

35 How to save the world

- The greenhouse gas problem
- Pacala and Socolow's wedges
- Engineering the climate
- Can we do it?

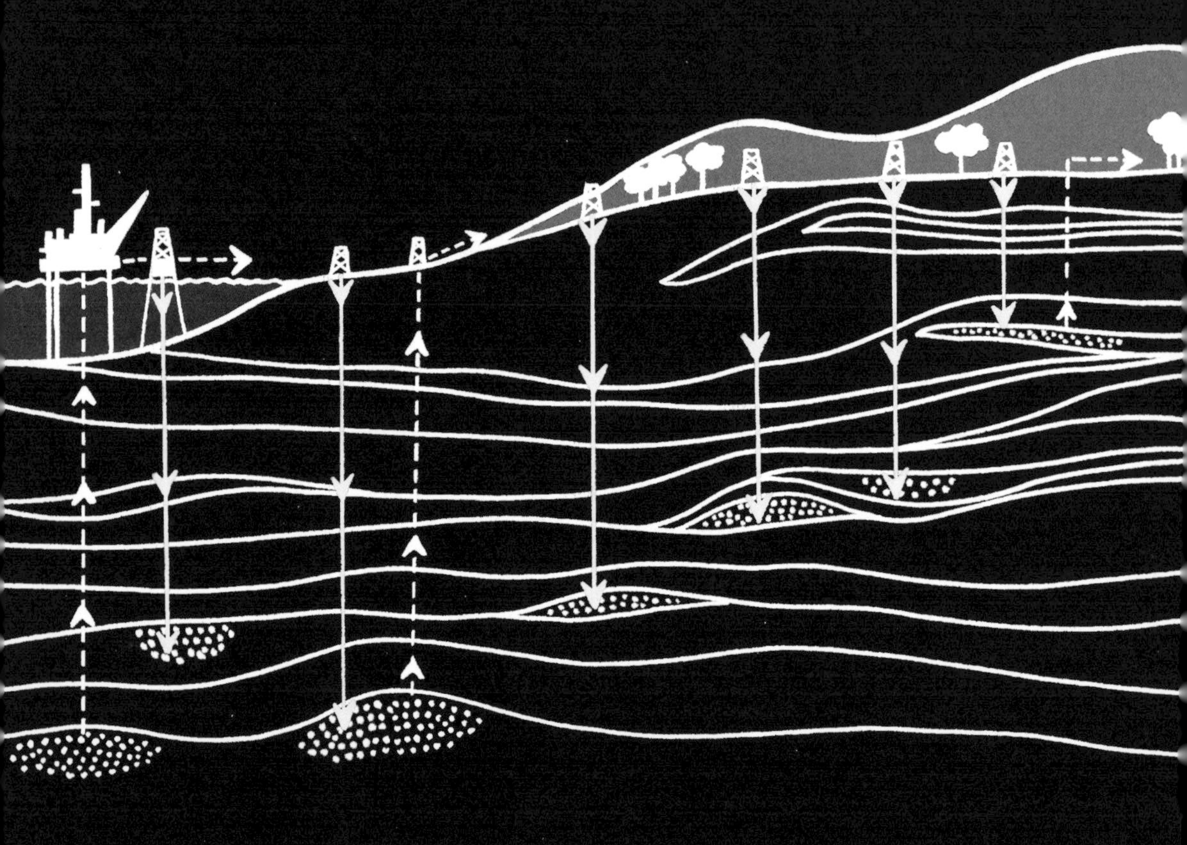

The world is heating up as the levels of greenhouse gases, particularly carbon dioxide, in the atmosphere rise. The exact details of how much, how fast and what areas will feel the biggest impacts of climate change are under continual revision, but there is no doubt what is happening. One response to climate change is to adapt how we live. Another path of action is to reduce the amount carbon dioxide released into the atmosphere. If that isn't enough to halt warming, it may be necessary to tinker directly with the climate itself.

The greenhouse gas problem

The current level of carbon dioxide in the atmosphere is around 380 parts per million (ppm) by volume. It is rising by 2 ppm every year. Before the Industrial Revolution, the CO_2 level was around 280 ppm. Since then, the world has warmed by about 0.8°C (1.5°F). Many countries have set a target to ensure that global temperatures do not rise by more than 2°C (4°F)above pre-industrial levels, and this is roughly associated in climate models to CO_2 levels of 400–500 ppm. Is it possible to meet this ambitious target?

Princeton scientists Stephen Pacala and Robert Socolow believe that the technology to cut greenhouse gas emissions already exists. No one technology on its own will do the trick, but the target can be achieved by combining just a selected few of them. If the world and its emissions continues to grow without any interventions (called the 'business as usual' scenario), greenhouse gases in the atmosphere are projected to double by the middle of the 21st century. To prevent the worst effects of global warming, and stay within the 500 ppm ceiling for greenhouse gases by 2050, Pacala and Socolow argue that net global emissions should be frozen immediately.

Pacala and Socolow's wedges

Pacala and Socolow display the different emissions scenarios on a graph, where the gap between the gently rising 'business-as-usual' line and the flat 'stabilization' line is a triangle that shows the amount of greenhouse gases that have to be prevented from reaching the atmosphere in the next 50 years. The triangle is further divided into wedges, each representing a different strategy to reduce CO_2 emissions. They say that implementing just seven of the 15 strategies they consider would stabilize greenhouse gases by 2050.

The stabilization wedges presented by the Princeton scientists break down the massive problem of reducing greenhouse gases into relatively bite-sized chunks. Each one represents the difference between following or not

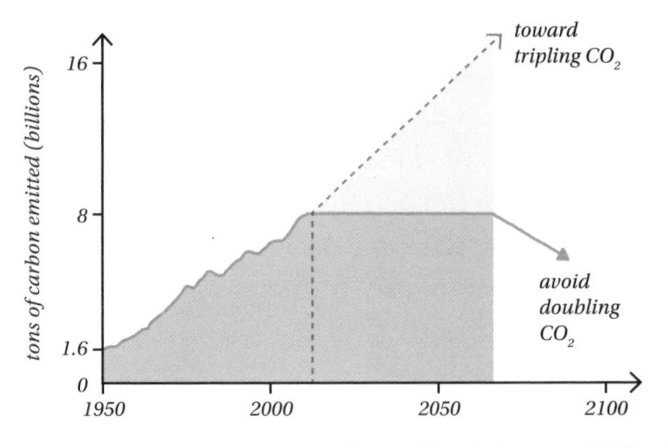

Pacala and Socolow's graph shows how carbon emissions have increased since 1950. The dotted line shows how emissions will increase in the future if current trends continue.

following a proven idea to reduce carbon emissions. They take in the whole range of technologies from energy efficiency and decarbonizing electricity and fuels to storing more carbon in forests and soils.

With vehicles, this means reducing the dependence on cars with better public transport systems and better urban design. Halving the annual distance travelled by 2 billion cars around the world would satisfy one stabilisation wedge. Another wedge could be achieved if the fuel efficiency of the 2 billion cars were doubled. Improving energy efficiency in buildings by a quarter would take care of another wedge.

Coal-fired power plants operate at around 32 percent efficiency and, in 2000, produced around a quarter of the world's carbon emissions. Assuming that there are double the number of coal-fired power stations in the world by the middle of the century, a stabilisation wedge would be satisfied if the stations could be made to operate with 60 percent efficiency.

Further wedges can come from replacing coal-fired power stations with relatively clean-burning natural gas or installing power stations with carbon capture and storage (CCS) technology. The International Energy Agency predicts the world's use of power will increase by 50 percent by 2030, with 77 percent of that coming from fossil fuels. CCS could prevent up to 90 percent of the carbon emissions escaping into the atmosphere by trapping and burying the greenhouse gases in the exhaust.

Doubling the number of nuclear power stations around the world would also take care of a wedge, as would a 50-fold increase in the amount of wind power or 2,000 gigawatts (700 times the capacity in 2004) of solar photovoltaic panels. Turning a sixth of the world's cropland over to growing sustainable biofuels, producing hydrogen (using renewable electricity to split water) to replace fossil fuels in cars or preventing deforestation would take care of a wedge each.

Pacala and Socolow stopped at 15 ideas but there are dozens more that could add up to produce the cuts in CO_2 emissions the world needs to make. Their point was not to prescribe a solution but, rather, to show how collective willpower could lead to significant results.

Engineering the climate

But what if that collective will to do restrict humanity's use of fossil fuels fails, as it seems to have done spectacularly in the United Nations climate summit in Copenhagen in 2009? Or what if the agreed reductions can only do part of the job? For more than 50 years, a fringe community of engineers and climate scientists have nursed thoughts of more radical ways to alter the climate. Many would no doubt dismiss the drastic ideas that fall under the umbrella of 'geoengineering' as reckless. They include creating artificial clouds to reflect away sunlight, creating colossal blooms of oceanic algae to soak up CO_2 from the atmosphere or giant sunshades for Earth.

The potential for dramatic and beneficial change to the climate through engineering has, until now, been outweighed by the risk of unexpected side-effects in the complex climate system that could have global consequences. But, as mitigation technologies begin to seem burdensome to the world's growing economies, geoengineering is coming in from the cold. John Shepherd, a climate scientist at the University of Southampton, argues that, with governments failing to grasp the urgent need for measures to combat climate change, radical – and possibly dangerous – solutions must be seriously considered.

Some of the most extreme ideas for climate engineering involve reducing the sunlight falling on Earth's surface as a way to offset the increase in temperatures caused by greenhouse gas emissions. Ken Caldeira, a leading climate scientist based at the Carnegie Institution in Stanford, California, has calculated that reflecting just 2 percent of the Sun's light from the right places on Earth (mainly the Arctic) would be enough to counteract the warming effect from a doubling of carbon dioxide in the atmosphere.

There are various options available to store captured carbon in underground cavities, some of which were made when mining fossil fuels.

❶ Depleted oil and gas reservoirs
❷ Use of CO_2 in oil and gas recovery
❸ Deep saline formations off- and on-shore
❹ Use of CO_2 in coal bed methane recovery
❺ Deep unmineable coal seams
❻ Basalts, oil shales, cavities

'Intervening in our planet's systems carries huge risks, with winners and losers, and if we can't deliver political action on clean energy and efficiency then consensus on geoengineering is a fantasy.'

DOUG PARR

One approach is to insert clouds of tiny particles, such as sulphur dioxide, into the upper atmosphere, where they would scatter sunlight. Dispersing around 1 million tonnes of sulphur dioxide per year across 10 million km^2 (4 million sq miles) of the atmosphere would reflect away enough sunlight. If this idea sounds fantastical, bear in mind that such climate interference occurs naturally through volcanic eruptions. When Mount Pinatubo in the Philippines erupted in 1991, for example, global temperatures dropped by 0.5°C (1°F) the following year, as dust was thrown high into the air. The dust then travelled right around the world, blocking out some of the sunlight. Another idea involves building 300-tonne ships that could spray micrometre-sized drops of seawater into the air under stratocumulus clouds. Proposed by Stephen Salter of the University of Edinburgh, the technology would make clouds whiter so that they reflect more sunlight.

Further into the realms of the fantastic are ideas to use shiny spacecraft to block sunlight. Scientists have proposed launching a constellation of free-flying craft that would sit between the Sun and Earth, forming a cylindrical cloud around half Earth's diameter and 10 times longer. The cloud would divert just enough rays from the Sun to reduce sunlight on Earth by 2 percent. The cost, however, would be a dazzling $100 bn a year.

Although techniques to stop sunlight from warming Earth would reduce temperatures, they would not tackle the fundamental problem underlying climate change: that of increased CO_2 in the atmosphere. Another class of geoengineering techniques focuses on tackling this issue. The growth of marine algae and other phytoplankton captures vast quantities of CO_2 from the atmosphere, but their numbers are often limited by a lack of essential nutrients in the sea water around them. Adding these nutrients would cause increased blooms of such organisms, which would sink to the bottom of the ocean when they die, taking the carbon with them.

Sucking CO_2 from the air could also be done mechanically. Klaus Lackner of Columbia University has designed a device that could, if built to full scale, take up the CO_2 emissions of 15,000 cars every year. It would take only

250,000 such machines to remove as much CO_2 from the atmosphere as the world is currently pumping into it. The UK's Institute of Mechanical Engineers is a fan of the idea, calculating that 100,000 of these artificial trees could absorb around 60 percent of the UK's total annual CO_2 emissions of 556 megatonnes. They calculated that forests of artificial trees powered by renewable energy and located near depleted oil or gas fields, where the trapped CO_2 could be buried, would be thousands of times more efficient than planting trees over the same area.

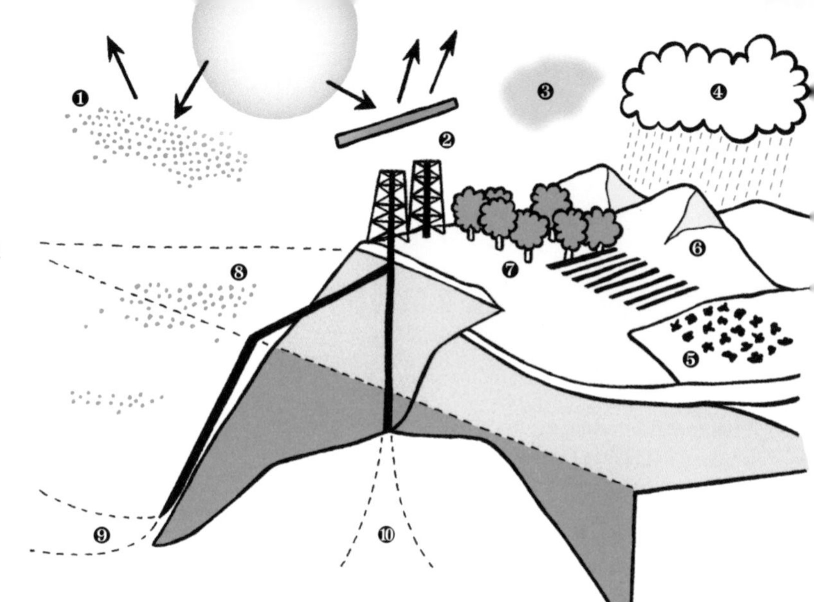

The major geoengineering schemes so far proposed:
❶ Aerosols in stratosphere
❷ Giant reflectors in orbit
❸ Chemicals to save ozone
❹ Cloud seeding
❺ Greening deserts
❻ Genetically engineered crops
❼ Grow trees
❽ Iron fertilization of sea
❾ Pump liquid CO_2 to deep sea
❿ Pump liquid CO_2 into rocks

Can we do it?

In the conclusion to its 2009 report looking at the desirability of geoengineering, the Royal Society said that, though they did not directly advocate geoengineering, experiments on some of the ideas had to start as soon as possible to ensure such mega-engineering plans are available as a safety net in case global talks to combat climate change failed. They considered capturing CO_2 from the atmosphere using artificial trees or shooting tiny particles into the upper atmosphere to reflect away sunlight to be the most promising ideas. But they stressed that all geoengineering techniques had major uncertainties regarding their own environmental impacts. No-one knows enough to make a decision one way or another.

In any case, technical and scientific issues may not be the stickiest problems when it comes to the deployment of geoengineering technology. Social, legal, ethical and political issues would be of equal significance and implementing global-scale projects would require international agreements. Critics wonder whether it is feasible to achieve the political stability and cooperation that will be needed to operate global-scale schemes. Ken Caldeira argues that all these issues need to be sorted out with urgency, given that we might soon have to confront the issue of whether, when and how to engineer a climate that is more to our liking. The worst situation is not to test the options and then react to a climate emergency by deploying an untested option, 'strapping on a parachute that you've never tested out as the plane is crashing'.

Glossary

absolute zero
Zero kelvin, the lowest possible temperature according to thermodynamics.

acid
A compound that, when dissolved in water, gives a solution of pH less than 7. Reacts with an alkali by donating hydrogen ions.

adenosine triphosphate
The molecule that transports energy from one part of a cell to another for use in metabolism.

alkali
A base that dissolves in water and has a pH of greater than 7. Reacts readily with acids.

atomic number
The total number of protons in an atomic nucleus.

atomic weight
The total number of protons and neutrons in an atomic nucleus.

axon
A projection from a nerve cell (neuron) that conducts the electrical impulses used to communicate with other parts of the body.

bacteria
A class of life that consists of unicellular microorganisms. They come in a variety of shapes, from rods to spheres and are one of the simplest life forms on Earth.

bit
Unit of computer information that can be a 1 or a 0.

bosons
Fundamental particles, such as photons and gluons, that carry the four fundamental forces.

Brownian motion
The random movement of small particles, as observed under a microscope. Named after the Scottish botanist Robert Brown.

butterfly effect
The idea, in chaos theory, that a tiny change in the starting conditions of a system can have a profound effect on its outcome. A flap of a butterfly's wing, for example, might change the course of a hurricane somewhere else in the world.

cell receptor
A site on the outside of a biological cell that allows hormones or other chemicals to enter or exit the cell.

cellulose
The primary structural component of the cell walls in green plants.

chlorophyll
A pigment found in plant and algae cells (and some bacteria), which captures the energy from sunlight to assist in the formation of sugars.

chloroplast
A part of the plant cell that contains chlorophyll and carries out photosynthesis.

chromosome
A length of DNA inside the nucleus of a biological cell, which contains a proportion of an organism's genes. Humans have 23 pairs of chromosomes.

cosmic inflation
The sudden expansion of the Universe in the moments after the Big Bang. The Universe more than doubled in size every 10^{-35} seconds and, by the time inflation was switched off at 10^{-32} seconds after the Big Bang, the cosmos had grown in size by at least 10^{43} times.

dendrite
The branched end of a nerve cell, which attaches to other nerve cells or parts of the body that need to communicate with the brain.

DNA
Deoxyribonucleic acid, the double-helix-shaped molecule that contains the instructions for making a life form.

entropy
The measure of the state of disorder in a system.

fermions
The class of fundamental particles that make up matter, including quarks and leptons.

fission
The act of splitting the nucleus of a heavy atom, which releases energy.

free radical
A highly-reactive by-product of the chemical reactions involved in metabolism. These are molecules with unbonded electrons and are extremely dangerous to body cells.

fundamental forces
The four forces of nature – electromagnetism, strong nuclear interaction, weak nuclear interaction and gravity.

fusion
The combination of two light nuclei (such as hydrogen) to make a new element, a process that releases energy.

galaxy supercluster
A grouping of several clusters of galaxies. The supercluster containing the Milky Way is 110 million light years across.

gene
The sequences of nucleotide bases that contain the codes for all the proteins in our bodies. Humans have around 25,000 of them.

genome
The full complement of DNA in an organism, comprising genes and non-coding DNA (previously known as junk DNA).

geoengineering
The large-scale implementation of measures to reduce the temperature of Earth. Ideas proposed have centred on reflecting the Sun's energy or taking greenhouse gases out of the atmosphere.

gluon
The particle that carries the strong nuclear force between quarks.

Higgs boson
Particle of the theoretical Higgs field, thought to confer mass onto other particles.

hormone
A chemical messenger in the body, used to control the metabolism of particular cells or organs.

ideal gas
A model of a gas that assumes all molecules act like billiard balls bounding around a volume.

ion
A charged atom, due to the presence of an extra electron (negative ion) or a lack of a full complement of electrons (positive ion).

ionization
The conversion of an atom or molecule into a charged particle by adding or taking away electrons or ions.

isotopes
Different types of atoms of a chemical element, where each one has a different number of neutrons.

kelvin (K)
Temperature scale where zero is defined as the lowest possible temperature achievable in the Universe. 0 K = –273.15 Celsius

kinetic energy
The energy of motion.

leptons
A class of matter particles including electrons and neutrinos.

Lorentz transform
The mathematical function, named after Dutch physicist Hendrik Lorentz, that allows observers to calculate how length and time change between different frames of reference.

metabolism
The full set of chemical reactions that govern the processes of life in an organism.

metamaterials
Artificial materials that have properties not seen in nature, such as negative refractive index.

neuron
A cell in the nervous system, either in the brain or further out in the body.

neutrino
Virtually massless, neutrally-charged elementary particle that travels almost at the speed of light and passes right through normal matter.

nucleotide base
The letters of the human genome – C, G, A and T – that form a sequence along a DNA molecule. The precise sequence of bases determines which protein a molecule will code for.

pathogen
An infectious agent that causes disease when it invades a host animal or plant.

pH
The measure of how acidic or alkaline a solution is. A low pH means a high concentration of hydrogen ions, while a high pH indicates a low concentration.

photoelectric effect
The phenomenon, explained by Albert Einstein, whereby metals emit electrons when illuminated by light of particular frequencies.

photon
A particle of light, with no mass, which also carries the electromagnetic force between particles.

prion
A non-living infectious agent that causes proteins to mis-fold in the body. Prions cause diseases such as bovine spongiform encephalopathy (BSE) in cattle and Creutzfeldt–Jakob disease in humans.

protein
Complex polymer made from amino acids. Proteins are the basic materials from which all life is made. They are also involved in carrying messages around the body and regulating its functions.

quark
The fundamental constituent of protons and neutrons. Quarks come in six flavours: up, down, top, bottom, charm and strange.

radioactivity
The emission of radiation from the nucleus of an atom. It can come in the form of alpha particles, beta radiation or gamma rays.

reproductive cloning
The act of somatic cell nuclear transfer (cloning) followed by letting the resulting embryo get to full term to produce a living organism.

RNA
A single-stranded version of the DNA molecule. RNA is used in body cells for multiple functions including reading the DNA code and creating proteins from the instructions.

senescence
The change in a biological organism after maturity. It may imply loss of function with old age or a change in the way genes are switched on or expressed.

sexual reproduction
The combination of parts of the genome from two parents to create offspring. This form of reproduction leads to greater genetic diversity in a population than other forms of reproduction, such as asexual.

somatic cell nuclear transfer / cloning
The technique for cloning animals, where the DNA from a donor's body cell is implanted into a hollowed-out egg, which is then implanted into a surrogate womb and grown into an embryo.

Standard Model of physics
A quantum mechanical description of all the matter and force particles that make up the Universe. Contains all the quarks, leptons and bosons.

statins
Drugs that can reduce the level of cholesterol in the blood.

stem cells
The body's master cells that, depending on how primitive they are, can turn into any tissue type in the body.

subduction
A process at the boundary of tectonic plats whereby one plate moves underneath another.

supernova
The huge explosion at the end of a star's life, which can outshine an entire galaxy for a short period of time.

telomere
The protective cap at the end of chromosomes that gets shorter every time a biological cell reproduces. When it is too short, the cell cannot reproduce any more. This is thought to prevent the occurrence of cancers and the build-up of dangerous mutations in the DNA of a living organism.

tesla (T)
The unit of measurement of a magnetic field. Earth's field is around 31 microteslas; a brain-scanning magnet is around 10 teslas.

theory of everything
The mathematical union of all four fundamental forces of nature into one framework. String theory is a candidate for the theory of everything.

therapeutic cloning
Exactly the same technique as carrying out reproductive cloning, but the embryo is not implanted and does not result in offspring.

thermodynamics
The laws of nature that govern the conversion of heat into mechanical work.

valence
A measure of the number of chemical bonds that an element can form. Usually related to the number of electrons available in an atom's outer shell.

virus
An infectious particle made from DNA or RNA inside a protein coat that enters living cells and uses the machinery there to reproduce itself. After a cell has reproduced the virus, it usually dies.

Index

Acknowledgements

Quercus Publishing plc
21 Bloomsbury Square
London, WC1A 2NS

First published in 2011

A catalogue record of this book is available from the British Library

Created for Quercus by Tall Tree Books Ltd

Senior editor: Rob Colson, Designer: Ben Ruocco, Indexer: Christine Bernstein
Cartoons and illustrations by Guy Harvey and Nathan Martin

ISBN: 978-1-84916-482-5

Printed and bound in China

10 9 8 7 6 5 4 3 2 1

Author's Note:

This book would not have been possible without the years of inspiration and encouragement from my colleagues at *The Guardian*. Many of the ideas in this book are partly or entirely based on the conversations I've had over the years with this remarkable group of people. So, thank you Emily Wilson, Ian Sample, David Adam, James Randerson, Tim Radford, Sarah Boseley Juliette Jowit, John Vidal, Simon Rogers, Clare Margetson, Mike Herd, Ben Goldacre, Robin McKie, Nick Hopkins, Ian Katz, Alan Rusbridger, Janine Gibson, Chris Elliott, Nell Boase and James Kingsland.

If the matchless Ruth Francis at *Nature* had not allowed me free rein to the archives of the world's best science journal, this book would have been a much poorer thing. Wherever I've drawn inspiration from scientific papers or books, I have done my best to credit the relevant people in the chapters themselves. But the following people deserve particular thanks for their unerring ability to explain science so clearly: Geoff Brumfiel, Kate Ravilious, Richard Fortey, Roger Highfield, Peter Atkins, Paul Davies, Michio Kaku, Phillip Ball, Michael Brooks, Marcus du Sautoy, Jim al-Khalili, Vivienne Parry, Bill McGuire, David Bodanis, John Ellis, Jenny Hogan, Ian Stewart, Mark Lynas, Robert L. Park, Laura Spinney, John Emsley, John Gribbin and Anil Ananthaswamy. I'm indebted also to the editors and writers of *Scientific American* magazine for key insights and explanations of some of the harder concepts of this book.